Clavius

Maginus

Longomontanus

Tycho

Walter

Deslandres

Regiomontanus

Pitatus

Purbach

Straight Wall

Mare
Nubium

Mare
Humorum

Arzachel

Alphonsus

Gassendi

Albategnius

Ptolemaeus

Grimaldi

Eratosthenes

Copernicus

Kepler

The
Apennines

Oceanus Procellarum

Timocharis

Archimedes

Aristarchus

Seleucus

Mare Imbrium

Sinus
Iridum

Plato

Mare Frigoris

Anaxagoras

PICTORIAL GUIDE
TO THE MOON

PICTORIAL
GUIDE TO
THE MOON

3rd Revised Edition

DINSMORE ALTER

Revised by
Joseph H. Jackson

Thomas Y. Crowell Company

New York   Established 1834

Title Page:

View of the lunar Oceanus Procellarum made from an altitude of 51 km by Lunar Orbiter II on November 25, 1966, showing the Marius Hills stretching below the crater Marius (upper right), a system of wrinkle ridges along the floor of Procellarum (center) and many of the rounded domes (lower right and center) probably caused by forces within the moon.

National Aeronautics and Space Administration

ACKNOWLEDGMENTS
Lick Observatory, Mount Wilson and Palomar Observatories (See Appendix)
National Aeronautics and Space Administration National Space Science Data Center
Jet Propulsion Laboratory, California Institute of Technology
A. A. Mills, Department of Geology, University of Leicester (Plates 16–6 and 16–8)
S. H. Zisk and T. Hagfors, *Radar Atlas of the Moon*: Final Report, NASA Contract NAS 9–7830; MIT Lincoln Laboratory, September 1970 (Plate 17–1)
G. V. Latham and M. Ewing, *Science*, Vol. 174, November 12, 1971, Figure 1, Page 687 (Figure 20)
W. L. Sjogren and W. R. Wollenhaupt, *Science*, Vol. 179, January 19, 1973, Figure 2, Page 277 (Figure 21)

Designed by Laurel Wagner

Manufactured in the United States of America

ISBN 0-690-00096-0

1 2 3 4 5 6 7 8 9 10

Library of Congress Cataloging in Publication Data

Alter, Dinsmore, 1888-1968.
  Pictorial guide to the moon.

  1. Moon—Photographs.   I. Jackson, Joseph Hollister.   II. Title.
QB595.A56   1973   523.3'9   73-9869
ISBN 0-690-00096-0

PREFACE

Whatever one's interest in the moon is, this book has much to offer. You may be one of those enthralled in the beauty of the moon as it rises huge above the horizon, sails high in the sky, and changes its shape and position through the month. Whether you view the moon through binoculars or a telescope, you may be intrigued with its stark contrasts and variety of features, or fascinated with the mysteries these have presented to scientists. As an amateur or professional astronomer, you may want to locate certain features and relate them to others on the full face of the moon.

All of these interests can be satisfied with the text, the maps, and the many brilliant photographs Dinsmore Alter made and collected for this book. Many of the photographs were made from earth, and, with this revision, others from lunar orbiters near the moon or from the lunar surface itself by astronauts. Dr. Alter's map of the moon in the astronomical convention and detailed maps in the astronautical convention based on closeup lunar photographs are all included.

One of the few astronomers who kept the interest in the moon alive with his enthusiasm a generation ago, Dinsmore Alter investigated its features thoroughly, as you can see herein. The odd obscurations he viewed and photographed over such craters as Linné and Alphonsus in the 1950's at first made little impression on the scientific world. In a few years, however, supporting evidence of brief patches and flashes came from the Soviet astronomer, N. A. Kozyrev, and from several American astronomers in Flagstaff, Arizona. Dr. Alter's analysis of these "transient phenomena" and of many other noteworthy features of the moon have proved remarkably penetrating and farsighted.

Dinsmore Alter died in 1968, after revising this book in 1967. Since that time, American-manned surface explorations and Soviet-unmanned vehicle probing of the moon have taken place. In the current revision, the latest opinions about the moon have been added, based on the direct lunar surface explorations, and many significant photographs from these missions have been included. But with this revision, the attempt has been made to hold to Dr. Alter's own goal, as stated in his original Preface, "to present data which may assist students of physical science toward decisions regarding which of the current and maybe contradictory hypotheses may more closely resemble the truth. . . . Today any dogmatism about these 'guesses' would be foolish. At any time new data may be obtained which will change a man's opinion. A scientist must be loyal to data and subordinate his most cherished beliefs to them."

CONTENTS

1 EARLY OBSERVATIONS OF

THE LUNAR SURFACE

With the many manned and unmanned landings on the moon, the era of its physical exploration has begun. Signals from instruments left on the moon and rocks returned to earth enable scientists to continue their study of the moon, which is just the latest chapter in the age-old story of man's relation to his environment. For the moon has always been there, hanging disarmingly close in the clear sky, dominating man's consciousness almost as powerfully as the sun. Its luminous presence, its beauty, and its manifest regularity of movement made the moon loom large in mythology long before it became the object of man's first efforts to observe and understand worlds beyond his own. In ancient legend, and even today in primitive parts of the world, the moon has been feared, loved, revered, and worshiped.

Sometimes the moon was considered a woman, perhaps the mother or the wife of the sun. Or it was a man or a male spirit, perhaps so potent in its maleness that a young woman could become pregnant from drinking the moonlit water upon which the lusty god had cast his light, as some of the Eskimos thought. Or the moon and the sun were ancestors of the human race, or relatives in perpetual conflict. Sometimes the legends became almost lyric in their poetic impulse, as when the Peruvian Indians conceived that the moon had given the gift of silver to the race of man in the form of its tears. All such legends had a common basis in mystery: the moon was intriguing precisely because it was quite unknowable. Centuries passed before the curious and rational Greek mind began to consider the movements of the heavenly bodies and attempted to reconcile the solar year with a year made up of lunar months. Thales of Miletus, the first of the Greek philosophers, is said to have produced the first important calendar reform around 600 B.C. His successors worked out various ways of harmonizing the solar and lunar cycles, but it was not until 45 B.C. that Sosigenes, upon the order of Julius Caesar, stabilized the year on the basis of the solar cycle alone.

It was the Greeks who discovered that the moon is not itself luminous but shines by the

1

reflected light of the sun. And one of the greatest of the Greek scientists, Aristarchus of Samos, made scientific history in the third century B.C. when he combined geometry, ingenuity, and some necessarily crude observations to determine the relative distances from the earth to the sun and the moon, and the relative sizes of the sun, moon, and earth. This was the first scientific attempt to measure astronomical distances; for the first time man began to understand the physical insignificance of his world. Aristarchus concluded that the sun was twenty times as far from the earth as the moon, a value that must have seemed wildly improbable to him, although the actual distance to the sun is 400 times as far as to the moon. His determination of the moon's diameter was remarkably good; he found it to be one third the diameter of the earth, while the true figure is somewhat more than a quarter. He figured the solar diameter to be seven times that of the earth. The true figure is 108 times, but even his figure was an astounding discovery for his century.

Most of the Greeks believed that the earth was fixed in space, with the sun, moon, and planets moving around it. Pythagoras considered the idea. Aristarchus began to conceive and elaborate a theory of the earth's motion, but Aristotle opposed the idea, as Hipparchus did later. Much later still Ptolemy, in his great book, *The Almagest*, crystallized the idea of a fixed earth. The smothering influence of Aristotle and Ptolemy continued almost unabated through medieval times.

Despite the Greeks' remarkable application of scientific techniques to the study of the moon, the moon itself remained quite unknown in any detail—an object of ingenious and romantic speculation. Both the essential change in outlook and the necessary instruments were far in the future.

The third great phase of man's relationship to the moon did not appear until a radical revolution in human outlook was under way. Gradually as the medieval world evolved into the modern era, and the metaphysical interpretation of life yielded bit by bit to the humanistic and scientific interpretation, man came to believe that he could by his own effort understand and even influence his environment. Great new reservoirs of human energy and imagination were released,

and remarkable discoveries proliferated. The characteristic achievement of modern times, the scientific and mathematical analysis of physical nature, was exemplified by the work of Galileo in the seventeenth century. Galileo's contributions to astronomy were based on one of the most significant inventions in the history of science: the telescope. The origins of this device are lost in the obscurity that has frequently enveloped the most pivotal early inventions, but the name of Hans Lippersheim, a lens maker in Middleburg, Holland, is often associated with the discovery. Lippersheim, about the year 1608, found that by combining two lenses in proper positions he could make distant objects appear larger.

The news of this invention soon passed beyond the Alps. Its implications were immediately apparent to Galileo, who constructed his own telescope in Padua during the following year. On a night in June of 1609, a memorable date in the history of human endeavor, Galileo turned his crude instrument toward the moon and the planets. The next year he published, in *Siderius Nuncius*, the results of his first detailed lunar observations. Although his telescope was very simple and disclosed little detail, Galileo's book included a crude map of part of the moon. It also included his determination of the heights of the lunar mountains, calculated by the timing of the first and the last sunlight on their tops.

Figure 1. Galileo's map of the moon. The large crater is probably Copernicus.

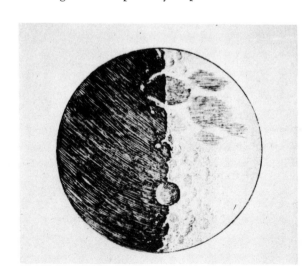

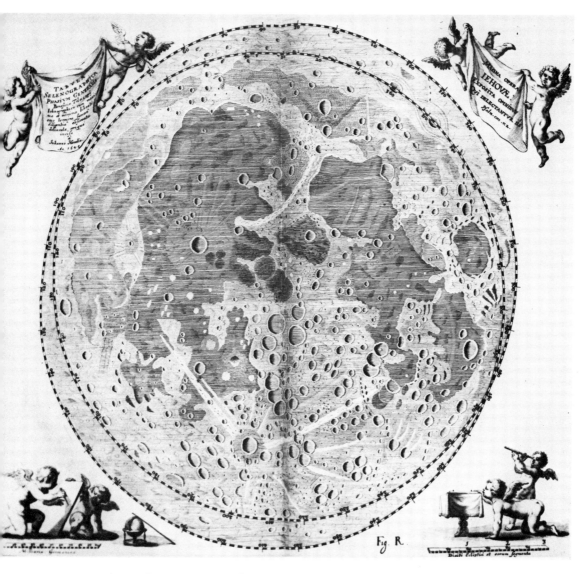

Figure 2. One of Hevelius' maps, from his classic *Selenographia*.

veryone who has seen a mountain peak bathed
sunshine and watched it for a time will re-
ember that the sunlight continues to flood the
p peaks for a time after the lower parts of
e mountain are in darkness. The same thing
appens on the moon, only the effect there is
ore conspicuous because of the lack of air.
'ith his telescope Galileo could see these lunar
ountain tops; the lighted peaks appeared al-
ost as stars close to the terminator, which is
e name given to the boundary between day
d night on the moon. He observed the time

when the starlike peak vanished and measured
the distance from it to the terminator. The calculation of height then became merely a problem
in trigonometry. (This method is reasonably satisfactory, although observational errors are impossible to avoid, even today, and they make the
resulting heights inaccurate.) The values Galileo
obtained were somewhat too large but still were
useful. Galileo also noted that because of the
tilt of the moon's axis to its orbit as it revolves,
an observer on the earth sees slightly different
parts of the moon's surface at different times.

This phenomenon, called libration in latitude, is discussed in Chapter 3.

Observations similar to those of Galileo were made at about the same time by his rival Christoph Scheiner, a professor at the University of Ingolstadt. His work was published in *Rosa Ursina* in 1626, but the book was devoted principally to solar observations and contained no lunar maps. Between 1620 and 1640 Langrenus, a cartographer to the King of Spain, mapped the moon and named some of its features. He is said to have engraved thirty maps of parts of the moon, but the names he used were never widely accepted and his work attracted less attention than it deserved.

Hevelius, whose long and distinguished ca

Figure 3. A map from Riccioli's two-volume work.

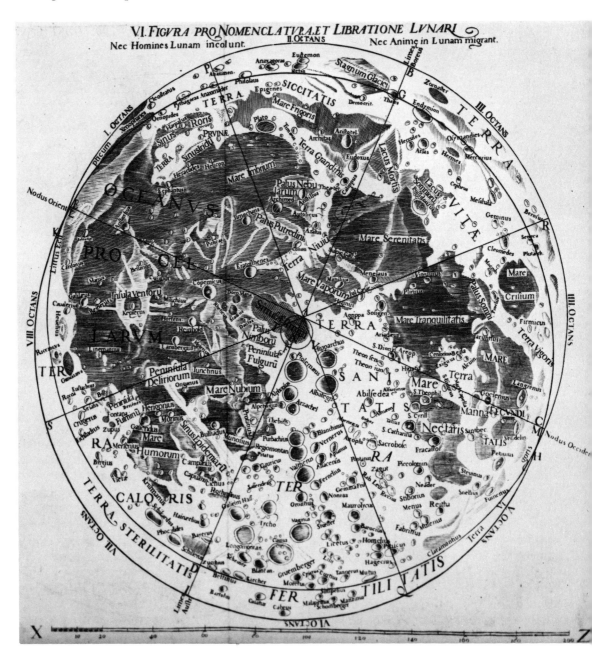

eer spanned much of the seventeenth century, published his great *Selenographia* at Danzig in 647. In it he rejected the names given by Lanrenus and proposed others on the basis of asumed terrestrial similarities. Except for six of his designations, his nomenclature was not permanent, but his maps were better than any produced for another century. One of these is reproduced here.

It will be noted that on this map north is printed at the top, just as on terrestrial maps, instead of at the bottom as on more modern lunar maps. This was the custom in the early days of moon mapping because the first telescopes used the type of eyepiece devised by Galileo, which produces an image right side up. In later and more efficient telescopes additional lenses would be required to produce an erect image, entailing a loss of light as well as extra expense. Since it is immaterial to the astronomer whether the image is erect or inverted, the extra lenses have been omitted and the resulting maps conform to the inverted image. Now that man is actually exploring the moon, the erect image with north at the top, just as in maps of the earth, is being used for lunar maps again. (Chapter 19.)

The maps of Hevelius show a great deal of detail very accurately. Almost uniquely among early maps, they show the areas of the moon's surface that are visible only part of the time because of the libration in latitude. The construction of these maps required many hours of careful observation with awkwardly mounted telescopes. The highest magnification Hevelius used was about forty diameters, and he estimated positions merely by eye. He discovered the moon's libration in longitude, the additional area seen at the east or west of the moon, although he gave an inaccurate explanation of its cause. The periods of the moon's rotation about its axis and of its revolution about the earth are equal while the rate of rotation is constant. The rate of revolution varies during the month because of the eccentricity of the moon's orbit. As a result sometimes the rotation gets ahead of the revolution but lags at other parts of the orbit. The resulting libration in longitude has a maximum value of a little less than eight degrees. Hevelius also measured the heights of some of the lunar mountains, obtaining more accurate values than Galileo had earlier by the same method.

Four years later Hevelius' friend, Riccioli, published a two-volume work. Probably many of the observations used by him were actually made by his associate at the University of Bologna, Grimaldi. Although Hevelius' work is generally rated as the more important of the two, Riccioli's is in some ways superior. It was he who set the pattern still followed in giving names to lunar features, and in the process he demonstrated a good insight into human nature. Although most of the craters were named after great scientists and other historic figures, Riccioli also honored politicians and others of his own generation whom he wanted to please. One of his maps is illustrated here.

Although astronomers following Riccioli performed much valuable work on the moon's orbit and librations and its relations to the sun and the stars, no outstanding cartography was produced for 125 years, aside from the excellent map of Cassini in 1692. In 1775 there was published posthumously the lunar map of Tobias Mayer, a professional map maker at the University of Göttingen. This work was the best of its kind for half a century: as late as 1837 Mayer's micrometric measurements of the positions and sizes of lunar markings were considered authoritative and useful. Mayer's chief fame, however, came from his investigations of the lunar orbit.

J. H. Schroeter published his first volume of lunar studies and observations in 1791 after nearly 30 years of systematic observations with reflecting telescopes made by Herschel and by Schraeder. Magnification ranged from 150 to 300 diameters. The largest of these, with a mirror 19 inches in diameter, was a truly impressive instrument for that time. Instead of using the micrometer to record the positions of lunar objects, Schroeter worked with the less accurate but more rapid "camera lucida." This consists of a special eyepiece for the telescope with a white screen divided into squares mounted below it. The observer looks with one eye through the telescope and with the other at the white sheet; the brain superimposes the two images, and one can trace rapidly what is visible in the instrument.

Schroeter was the discoverer of what he called "rills," and are now often called "clefts." These are long, narrow valleys in the surface, some of them more than a hundred miles long. About 2,000 of them have been mapped to date.

Schroeter believed that he had discovered evidence, through twilight observations, of a lunar atmosphere, a subject which is considered in detail in Chapter 2. He was also the first to introduce a relative scale for measuring the brightness of lunar features.

Schroeter was followed by two Germans, Wilhelm Beer, a banker, and Johann Heinrich von Mädler, a professional astronomer. Their *Mappa Selenographica* and their book *Der Mond* were published in 1837, and their description of the moon was the basic reference work for many years. It is still of great interest to professional astronomers and amateur lunar specialists. Such subjects as eclipses, the moon's path, and its rotation are discussed at the beginning. There are long tables of measurements of the latitude and longitude of lunar markings. Lists of various names given to the same features are included, as are determinations of the heights of many of the lunar mountains. Today the uncertainties in these measurements are understood, but the data were accepted for a long time and are still some-

times quoted. The authors presented evidence (which is no longer accepted) for the existence of a thin lunar atmosphere. A special chapter signed by Mädler, considered lunar effects on our terrestrial weather. Although later scientists considered such effects a myth, studies during the 1960's pointed to these effects again, and the question has not yet been settled. The bulk of the book described lunar features in detail.

Almost forty years later, when Edmund Neison published his comprehensive treatise in selenography, he referred to *Der Mond* as "the only work on the subject" and predicted that for many years to come all who followed his distinguished predecessors would have to incorporate the results of Mädler's seven years of observations. But most of the material in Neison's book was new, the result of eight years of constant selenographic observation. This work was done mostly with a six-inch equatorial telescope and included several hundred sketches sent to the author by other observers, notably by the Reverend T. W. Webb. Neison defended the thesis of a lunar atmosphere very strongly, associating it with the phenomenon of weathering and commenting that "the entire evidence we possess on this subject is strongly favourable to the moon actually possessing such an atmosphere."

Two years before Neison's book appeared there was published in England a quite different book by Nasmyth and Carpenter. It was much less important than Neison's work, but it is perhaps the most beautiful of the standard treatises. The illustrations are so excellent that the unique work of the authors in preparing them should be told in their own words, as an illustration of the highest kind of scientific devotion and craftsmanship:

"During upwards of thirty years of assiduous observation, every favourable opportunity has been seized to educate the eye not only in respect to comprehending the general character of the moon's surface, but also to examining minutely its marvelous details under every variety of phase, in the hope of rightly understanding their true nature as well as the causes which had produced them. This object was aided by making careful drawings of each portion or object when it was most favourably presented in the telescope. These drawings were again and again re-

Figure 4. Title page of *Der Mond*, by Wilhelm Beer and Johann Heinrich Mädler. Published more than a century ago, it is still of interest to selenologists.

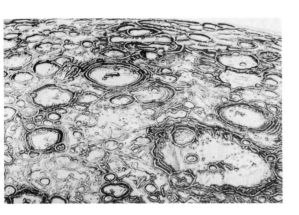

Figure 5. A section of a map by Schmidt of Athens. It shows Clavius.

peated, revised and compared with the actual objects, the eye thus advancing in correctness and power of appreciating minute details, while the hand was acquiring, by assiduous practice, the art of rendering correct representations of the objects in view. In order to present these illustrations with as near an approach as possible to the absolute integrity of the original objects, the idea occurred to us that by translating the drawings into models which when placed in the sun's rays, would faithfully reproduce the lunar effects of light and shadow, and then photographing the models so treated, we should produce most faithful representations of the original. The result was in every way highly satisfactory and has yielded pictures of the details of the lunar surface such as we feel every confidence in submitting to those of our readers who have made a special study of the subject."

Most of their book is devoted to the marshaling of arguments in favor of a special form of volcanic hypothesis. The authors made skillful use of the physics of their day, and although much of their discussion has not stood the test of later research, a book of this quality must never be ignored.

The best map of the moon made during the pre-photographic days was the 72-inch one by Julius Schmidt (Schmidt of Athens). The map was published in 1878 by the Prussian government. During the latter part of the nineteenth century, photography began to supplement visual map making although some work of the

visual type still persists usefully. The first fifty pages of Walter Goodacre's *The Moon*, published privately in 1931, contains an excellent historical outline.

Throughout the centuries there has been controversy over the proper names given to lunar features. As has been noted, names were first assigned by Langrenus. Then came Hevelius' nomenclature and a few years later Riccioli's. Beer and Mädler standardized the names and added many new ones. Neison adopted their list but added still more, for a total of 513 proper names. In 1878 Schmidt of Athens added about 50 more. In many cases, and particularly in the earlier years, favoritism and politics dictated some of the choices, to the credit of otherwise obscure persons and the detriment of famous scientists. In 1932 the International Astronomical Union standardized the nomenclature by formal acceptance of a report by Mary A. Blagg and K. Mueller.

This outline leads us to the "modern" era of lunar observations which may be considered (rather vaguely) as beginning with the systematic observations by Dr. J. H. Moore and Mr. J. F. Chappell, using the 36-inch refractor of the Lick Observatory as their camera. Some earlier photographs were fully as good as their best ones, but their systematic practice between the years 1937 and 1946 gives their series preeminence. It was the custom at the observatory on nights when the "seeing" was unusually steady and the moon was available, to call these observers. They made several exposures during an hour, then turned the telescope back to the regular observer of the night.

Among the selenographers worthy of consideration during the latter part of the nineteenth and the early part of the twentieth century are: W. R. Birt of England (1804–81); Moritz Loewy of France (1833–1907); Henry Draper of the United States (1837–82); T. G. Elger of England (1838–97); Hermann Klein of Germany (1844–1914); Julius Franz of Germany (1847–1913); S. A. Saunder of Cambridge, England (1852–1912); N. S. Shaler of Harvard (1841–1906); William H. Pickering of the United States (1858–1938); Johann Krieger of Germany (1865–1902). Some of these men were professional astronomers; others were amateurs, but all loved the moon.

7

2 GENERAL CONDITIONS AT THE MOON'S SURFACE

If the beachhead of the first lunar landing is to be enlarged, if humans are to use the moon wisely and successfully, man will have to accustom himself quickly to a wondrous new environment. There will be dangers without precedent in terrestrial life—and, surprisingly, certain factors of assurance and safety that are absent on the surface of the earth. But all of these things, the good and the bad alike, will be only details in the vast strangeness of an existence to which all the millennia of the human race afford no parallel.

THE DAY ON THE MOON

Life on the moon will completely disrupt one of the most profound and deeply ingrained of man's environmental rhythms—the alternation of day and night, of light and darkness, in a 24-hour cycle that governs work, eating, and sleep. On the moon wondering humans will observe that the circuit of the sun in the heavens, from one dawn to the next, will require the equivalent of 29.53 terrestrial solar days. The schedule of human activity will no longer be related to the external, astronomical clock to which all his existence has accustomed man. No one can predict with assurance what effect this will have on the human body or psychology. Perhaps man's physiology, reflecting the inheritance of unnumbered generations, will still demand the terrestrial working day. It may be that a cycle of shorter periods will better satisfy the needs of the organism. Perhaps, on the other hand, the human body will be adaptable to longer periods of work because of a number of environmental changes. For example, the weight of the body and hence the effort required for various exertions will be much less; the artificial atmosphere provided for human use will be different; food will probably have to differ considerably from the customary terrestrial diet.

Even the appearance of the heavens by day and by night, the progression of phenomena across the sky, will be strange. The sun and the

…rs will move very slowly in their paths. The …rth will be visible, great in size and in clear …tail, but its apparent motion will be either …ght or spectacular depending on the point on … moon's surface from which the observation …made. From the point that appears to us to … the mean center of the moon's disk, the earth …ways will appear very high above the horizon. …cillations in the earth's motion will appear to … at a minimum in this case, but for an observer …tioned at certain positions near the limb (the …ge as seen from the earth) of the moon, the …rth may rise and set, sometimes climbing al-…st 16 degrees in a very irregular manner, …ving sideways in one direction and then the …er before it begins its descent to the horizon. …om other points near the moon's limb the earth …ll barely rise above the horizon before dipping …ck out of sight. The earth will be seen rising …various points on the horizon, depending again … the position of the observer along the limb.

MPERATURE CHANGES

The moon's lack of air, which will require …n to provide his own atmosphere as a condi-…n of survival, will produce another dramatic …nsequence in strong contrast to terrestrial con-…ions. On the earth, weather is a complicated …oduct of very involved processes in which …nperature, winds, tides, and atmospheric con-…ions are the controlling factors. On the moon, …ather is mathematically predictable for any …signated time, even far in the future. The …ly relevant factors are the time of day, the …sition of the observer on the moon, the state … the moon's librations, and its distance from … sun; and all of these are ascertainable and …eseeable for any time and place.

As a matter of fact, the temperature of the …face rocks on the moon has already been …asured with moderate accuracy from the earth, …quarter of a million miles away. The chief …trument in the process is an extraordinary …vice so tiny that hundreds of them together …igh no more than a split pea. It is called a …rmocouple and is really nothing more than … very fine, short wires of different materials …ed together. The wires have been generally … bismuth and an alloy of bismuth and tin.

Any radiation—such as that contained in a beam of light—that falls on the junction of these tiny wires causes the wires to emit a very small but measurable electric current. The current can be measured by a galvanometer connected to the open ends of the wires; the size of the current gives the intensity of the heat contained in the light beam or other radiation.

The thermocouple cannot be enclosed in glass because glass is opaque to much of that part of the heat radiation which is in the infrared range. The radiation, therefore, usually is exposed to the junction of the thermocouple through a polished rock salt window. Surprisingly, the smaller the thermocouple, the better it performs. As its size is decreased it becomes more definitive, responding to the light from only a small part of the moon's disk, measuring more precisely the heat from a more exact source area. And the smaller the device, the greater is its surface area in respect to its mass; the radiation it receives therefore raises the junction to a higher temperature, increasing the electric current and permitting more accurate measurement.

This instrument was used with signal success by Pettit and Nicholson, and later by Pettit alone, with the 100-inch Hooker telescope at Mount Wilson and smaller telescopes. They followed the subsolar point (the point on the moon's surface with respect to which the sun is at the zenith) across the moon's disk during fifteen consecutive terrestrial nights. In the course of these experiments Pettit and Nicholson encountered several problems inherent in the nature of the radiation they were attempting to measure.

The first of these is that there is a tendency toward what is called specular radiation and reflection. That is, the surface of the moon behaves much as a convex mirror does in reflecting light: too much radiation comes toward us when the subsolar point is at the center of the visible disk, and too little when it is near the limb. The result is a wide variation in the electric current stimulated by the radiation, and this must be corrected for.

Second, the energy that reaches us from the moon's light consists of two parts, and these must be separated in order to make temperature determinations. One part is the radiation from lunar rocks heated by the sun; it is this energy that

9

the observer wants to isolate and measure. He must separate and eliminate the second part: the portion of the sun's radiation that is reflected directly from the moon to the earth instead of being absorbed and heating the moon. This second part is quite independent of the temperature of the reflecting rocks on the moon. Fortunately the radiation from the lunar rocks—the radiation we seek to measure—is in the infrared range, longer in wavelength than is nearly all of the radiation from the sun, and separation of the two components on the basis of wavelength is a purely mechanical procedure.

Finally Pettit and Nicholson found that the tables indicating the amounts of radiation absorbed by the earth's atmosphere, a phenomenon for which due allowance must be made, were incorrect. They were forced to make new tables based on their own experiments.

After struggling with these various areas of difficulty, Pettit and Nicholson finally determined that the moon's temperature at the subsolar point is 101° Centigrade, or 214° Fahrenheit. This result corresponded well with the theoretical value.

The temperature thus determined is subject, however, to very wide variation at other points on the moon's surface. Observations of the temperature near the limb, instead of at the subsolar point, have been made during total eclipse. These observations indicated a drop from 69° C (156° F) to −98° C (−146° F) during the first partial phase of the eclipse. During the time of total eclipse the temperature at the limb dropped slowly to −116° C (−177° F) but it rose to nearly the normal value before the eclipse had completely terminated. This great range of temperature in so short a time indicates that the moon absorbs very little of the radiation falling on it. Only a very thin surface layer is warmed materially. The temperature on the night side of the moon, which of course faces earth during the crescent phase, has been found to be −155° C (−247° F), but this value is subject to considerable uncertainty because the thermocouple produces only a tiny current at extremely low temperatures.

These measurements have recently been checked and confirmed by Sinton, working at the Lowell Observatory with more delicate sensors. He has established isotherms (lines joining points of equal temperature as on a conventional weather map) and has found, in addition certain variations in temperature that result from the nature of the surface in local areas.

The enormous range of temperature on the moon and the quick progression from one extreme to another will pose a large problem in self-protection for the first humans attempting to cope with these unprecedented conditions But there is at least one advantage: the high daytime temperature should make the problem of power supply almost unbelievably simple. The temperature of the subsolar point is just above the boiling point of water as we know it at the earth's surface. Water tubes, blackened to cause maximum absorption and mounted in cylindrical reflectors, may be exposed to the direct radiation of the sun for the equivalent of two terrestrial weeks without interruption. These tubes need merely be mounted on axes parallel to the moon's axis of rotation, and then turned at the proper speed to match the moon's rotation. Water or any other convenient fluid may be used—if the adjective "convenient" can apply in reference to a process that may require the transportation of all fluids through a quarter of a million miles of space! The steam produced by heating the water tubes will, of course, have to be repeatedly condensed and later reheated to conserve the precious fluid. The sun would thus act as a perpetual source of fuel for lunar steam engines practically no cost after initial installation—a possibility that does not, at least at present, exist on the earth.

Solar batteries can also be used advantageously in the clear, airless lunar sky to change solar radiation directly into electrical energy Although they are not yet capable of delivering heavy current, their development is barely in infancy, and rapid progress has already been achieved. It is certain that devices of this kind will have valuable uses in lunar scientific laboratories, and we can hope that they will prove be the best source of current for general use.

A third source of power is either atomic fission or fusion. Atomic engines would be needed at night when sources dependent on solar energy fail, unless adequate storage batteries can be developed.

The question of a lunar atmosphere was the ubject of lively discussion for 250 years, after drien Auzout pointed out, during the latter part the seventeenth century, that if the moon had n atmosphere there should be twilight. This ould be most easily visible as a faint light exnding a short distance into the unlighted part the moon. It was more than a century before y astronomer claimed to have observed such effect. Schroeter announced the discovery afr observing the moon as a very thin crescent, which time the effect should be most noticele. He determined that an atmosphere suffient to produce this effect would have to be a ifle more than one mile in height. The moon osses this distance in its orbit in less than two conds of time, making observation of any fect of atmosphere very difficult. Today hroeter's results are known to have been merely usory.

If there is a lunar atmosphere, the instant contact of a total solar eclipse should be anged because of the refractive effects of the mosphere. That is, an atmosphere would form outside envelope that would bend light and ange the moon's apparent size and the position its limb. Leonhard Euler observed the eclipse 1748 and decided that the moon does have an mosphere with a horizontal refraction of 20 conds. (The earth's atmosphere has an index 2 times as great.) Unfortunately Euler negcted the effects of irradiation. The French tronomer Du Sejour attempted to remedy this fect in Euler's reasoning at an eclipse sixteen ars later. The resulting refractive value was ly 0.5 second, which gave to the moon an mosphere with a density $\frac{1}{1,400}$ that of the rth's.

A star is said to be occulted when the moon sses between the earth and the star, tempo-rily cutting off the star's light as observed on e earth. At the moment such an occultation curs, the lunar atmosphere, if there were any, ould produce three effects: a change in the lor of the star, a dimming of the star and its sappearance from view a very brief time later an the predicted moment of invisibility, and re-emergence into view the same short time

ahead of schedule. The reason for these variations in the predicted times of disappearance and reappearance is, again, the bending, or refraction, of the star's light as it passes through the moon's atmosphere, so that it comes to our eye either later or earlier than it would if it had followed a straight line, unaffected by atmosphere. A few observations of such phenomena were reported long ago by capable, reputable astronomers. In general, however, these results were not verifiable and most astronomers of the nineteenth century reported that no observable effects existed. Even if we were to assume the accuracy of Du Sejour's eclipse results, it is questionable whether any such effects would be great enough to be noticeable at stellar occultations.

The great landmark in speculations concerning lunar atmosphere occurred in 1860, when the famous English physicist James Clerk Maxwell published his monumental *Kinetic Theory of Gases*. Maxwell's work established that the moon could possess no atmosphere of more than negligible density.

Maxwell assumed that a gas is composed of individual particles moving at random. The pressure of a gas on its container is determined by the number of particles that strike a given area in a given time, and by the average kinetic energy of these particles. This energy of motion is the heat contained by the gas. The speed of the particles of gas, therefore, varies with the temperature: at absolute zero the particles would not move at all; at low temperatures they move slowly and at high temperatures they move very fast. Maxwell applied the laws of probability to the random movements of gas particles and worked out the relationships between pressure, temperature, and the speed of the particles in his theory.

What happens in a mixed gas composed of various kinds of particles? As the particles dash about they collide with one another and rebound like millions of billiard balls in endless agitation. The collisions transfer energy from the more energetic particles to the less; the result is an "equipartition of energy," a strong tendency for all particles to have the same amount of energy. Since kinetic energy is a function of mass and the square of velocity, the lighter particles must move faster than the heavy ones.

Specifically, the velocities of the particles will be inversely proportional to the square roots of their masses: if one molecule has four times the mass of another, its speed will be only half as great.

At any one time some molecules of a given mass will have more energy (and thus higher speed) than the average of their kind and others will have less. But the tendency to approach equality makes it possible to determine the average velocities of the particles in a gas at any temperature. The following table shows the average speeds of some particles at 0° C (32° F) and at 100° C (212° F). The first part of the table lists some individual atoms. But in most gases the particles are composed of two or more atoms joined together, and the second section of the table gives the velocities of four such molecular gases.

Table 1

Element	Atomic weight	Speeds in km/sec at 0° C	100° C
Hydrogen	1	2.62	3.06
Helium	4	1.31	1.53
Carbon	12	0.76	0.88
Nitrogen	14	0.70	0.81
Oxygen	16	0.66	0.77
Neon	20	0.58	0.68
Argon	40	0.41	0.48
Molecules other than atomic			
H_2	2	1.84	2.15
H_2O	18	0.62	0.72
O_2	32	0.46	0.56
CO_2	44	0.39	0.46

These molecular velocities have a direct bearing on the question of a lunar atmosphere because an atmosphere must be held in place, so to speak, by a gravitational force. Now, a gas molecule moving directly away from the surface of the earth and not striking anything requires a velocity of 11.188 kilometers per second to escape completely from the earth. Under similar conditions on the moon a speed of only 2.38 kilometers would be required. A glance at the speeds shown in the table would suggest that only hydrogen could escape from the moon,

and that there should, therefore, be a considerable atmosphere of the other elements—even including hydrogen in combinations such as water vapor. But we must remember that the values in the table are average speeds. As a result of deviations from the averages, some molecules move quite fast enough to escape if they are not stopped by an encounter with other particles. In a very short time (as measured on the scale of planetary development) almost all of the lighter atoms in any lunar atmosphere would have escaped into space. In any event, the moon can hardly have had much of an atmosphere any period of its existence.

The heavier of the chemically inactive elements, such as xenon and krypton, exist only a very small percentage of the terrestrial atmosphere; a lunar atmosphere composed exclusively of them would have an extremely low density. However, unless there be outside interference, such an atmosphere must exist.

Recent observations of space conditions near the earth's orbit, made from artificial satellites, have revealed the presence of streams of very rapidly moving protons that have been termed a "solar wind." Mariner II, in December 196 has shown this wind to exist far from us. If the moon had a moderately strong magnetic field as does the earth, these ionized particles would have little effect on any lunar atmosphere. The lunar surface would be protected from them. But, although observations to date have not given positive proof, it appears that if the moon has a magnetic field at all it is very weak. If this is true the "solar wind" sweeping past the lunar surface during eons of time, striking atmospheric molecules and accelerating their motion, would surely have stripped the moon almost bare. This stripping effect is indicated also by radar measurements of the density of the lunar atmosphere, which indicate a measured density of 10^- times the surface density of the earth's atmosphere. In other words, the density of our atmosphere, according to this research, is ten trillion times that of the moon! Even allowing for the maximum error we can conceive, a lunar atmosphere in the ordinary sense of the word must be almost nonexistent.

Yet there remain some bothersome observations tending to suggest an "atmosphere" effect. Three centuries ago, Hevelius reported that arkings on surface features of the moon were ometimes less bright and less regularly defined han at other times, although the earth's sky was ist as clear. No thin atmosphere such as is onceivable on the moon could possibly account or such obscuration. But similar observations ontinue to be made in certain lunar areas and t certain times of the lunar day. They are too ell authenticated to be ignored.

These observations are perhaps best conrmed within the crater or ringed plain known as 'lato, and on the left part of the floor of the alled plain Alphonsus. Goodacre sums up the bservations of Plato under three headings.

1. The darkening of the floor from the time he sun is 20° above the horizon until afternoon. ctually the floor does brighten, but to a lesser xtent than do the features around Plato.

2. Variations in the visibility of the light pots and streaks without relation to the solar ltitude.

3. Occasional obscuration of easily observed eatures as though by fog. This takes place even hough adjoining parts of the moon are seen learly. Neison reported an observation in which he whole interior was obscured by fog at sunise. Klein observed a fog-like effect in 1878. . Hodge, observing Plato near sunset, could find o trace of craterlets. Goodacre confirmed this bscuration by his own observations three hours ter. These are merely a few samples of such bservations.

There is nothing surprising about such reorts of what appears to be a local haze. There no known reason why there should not be some esidual volcanic activity involving the leakage f very low density gases from some of the lunar raterlets. Such gases could have an obscuring ffect at the moment of their escape although ney could not form a permanent atmosphere. ecent observations have suggested that some uch very slight effect is quite probable. (The natter is discussed in some detail in Chapter 6.)

HANGES ON THE MOON

From time to time observers have reported changes" on the surface of the moon. Changes f various kinds are certainly theoretically possi-

ble, but it is questionable whether, during the past two centuries, permanent changes large enough to be observed from the earth have occurred. ("Fogs" of the kind already described are an example of temporary changes.)

The moon is certainly bombarded by meteorites, just as the earth is. But not even a meteorite crater on the moon as large as the famous one in Arizona could be observed in any detail from earth. Slippages from steep cliffs, triggered by meteoritic bombardment, may have occurred, but, again, probably not on a scale large enough for terrestrial observation during the short time we have had telescopes.

Optical effects probably explain most of the "changes." Everyone who has done a modicum of lunar observing is aware of complete changes of appearance with the altitude and direction of the sun, even small changes in altitude sometimes producing marked variations. A return to the same phase of the moon may not necessarily bring the sun to the same altitude. Librations vary the altitude by a dozen degrees. In addition to this difficulty, many suspected changes are based on old observations. Beyond this there is a psychological factor that influences everybody; it is easy to accept the exciting explanation instead of the prosaic but probable one.

One should certainly keep an open mind regarding permanent observed changes, but the writer is very skeptical about them, except perhaps for residual volcanic activity in the craterlet Linné. Even there it appears probable that varying amounts of haze have produced the observed data. Any reader who wishes to pursue the subject in more detail should read the monograph by Walter H. Haas, who is a very capable and enthusiastic student of the surface of the moon. Appended to it is a bibliography of 83 references.

ALBEDO

The albedo is the ratio between the light which the moon reflects and the light that falls on it. The moon reflects from ordinary parts of its surface only about one-fourteenth of the incident visible light; in other words its albedo is low. A minor cause of this low albedo is the roughness of the surface. The most important cause is the fact that the greater part of the surface is composed of rather dully colored material. There is, however, a very great variation in the albedo of the different features. This is shown conspicuously by the fact that the waning moon, just before the half moon phase, gives us about 20% less light than does the corresponding phase of the waxing moon. The obvious reason is that the lighted area of the waning moon happens to include a greater proportion of the dark plains, called maria. See Plates 3–1 and 3–2.

The angle between the general surface and the line from us to that surface also affects the amount of light we receive. Hence, the albedo cannot be determined at the full moon phase alone because of the tendency toward specular, or mirror-like, reflection. If the moon were a smooth, perfectly polished globe we would receive light from only one spot. Even as it is, the brightness increases very rapidly during the couple of days preceding full moon and falls off very rapidly after that phase has been passed. Although the illuminated area turned toward us at the half moon is half as great as at the full moon, we receive only about a ninth as much light, primarily because of this mirror effect.

In addition to albedo and phase, the distance of the moon from the earth helps determine the amount of light we receive. As an object is moved farther away the angle it subtends (i.e., its angular diameter) is reduced. The angular area varies with the square of the distance. When the moon is closest to us it is only 221,000 miles away; half a month later its distance is 253,000 miles. On account of this variation of distance, one full moon may be 30% brighter than another. Visually, the full moon gives us, on the average, a little more than $\frac{1}{500,000}$ the light we receive from the sun. Photographically, it is even less bright, because of the general slight yellowish tinge of its surface. Part of this tinge, however, is spurious and is due to two non-lunar factors. The first of these is the effect of the earth's atmosphere, which, in screening the light coming through it, absorbs the short, bluish wavelengths more effectively than it does the longer, reddish ones. The second is the greater sensitivity of the normal human eye to the medium (yellow) wavelengths than to longer or shorter wavelengths.

SCALE OF BRIGHTNESS

The albedo of the moon varies enormous. from feature to feature. Schroeter devised a sca from 0 to 10 to record the brightness. This wa somewhat modified by later observers, and esp cially by Beer and Mädler. The scale is now de fined as follows:

Table 2

0	Black—the dark shadows
1	Grayish black
2	Dark gray
3	Medium gray
4	Fairly light or yellowish gray
5	Light gray
6	Approaching grayish white
7	Grayish white
8	White
9	Glittering white
10	Dazzling white

The interpretation and application of such scale is obviously difficult for even a moderate experienced observer. Many areas blaze out ne full moon and appear of entirely different brigh ness relative to other parts. Perhaps the best trai ing for the serious student would be to rea Neison's description of each feature and to con pare it with his own observations.

The following table of brightnesses is take from Neison. (When "walls" or "interior" mentioned, the reader must not assume that the are no parts of these which are not much bright or darker than the average values stated.)

Table 3

Feature	Brightne.
Maria	
Imbrium	
general	$2\frac{1}{4}$ to $3\frac{1}{}$
Sinus Iridum	$2\frac{1}{4}$
streaks	4 to 5
Nubium	2 to 3
Oceanus Procellarum	2 to $3\frac{1}{}$
Serenitatis	
central	3 to $3\frac{1}{}$
border	$1\frac{1}{2}$ to $2\frac{1}{}$
Tranquillitatis	$1\frac{1}{2}$ to 3

Feature	Brightness	Feature	Brightness
aters, Ringed Plains, etc.		Kepler	
Alphonsus		walls	7
central peak	7	floor	6
Anaxagoras		Linné	5½
wall and interior	7	Manilius	
Archimedes		wall	8
floor, dark parts	2½	interior	4
light parts	3½	Philolaus	7
east wall	5	Plato	
Aristarchus		walls	6
central mountain	10	floor near	
interior, craterlet and		half moon phase	3
second peak	9½	just after full moon	1+
Aristillus		Posidonius	
interior	3	general interior	3½
Aristoteles		Ptolemaeus	3½
walls and floor	4	principal interior crater	7
Arzachel		Riccioli	
walls	4	floor	1 to 1½
Biot	8	Timocharis	4
Copernicus		Tycho	
floor	3 to 4	general interior	5
slopes	5 to 6	central peak	8
crest	8	base of exterior slope	3
certain peaks	9		
Eudoxus			
interior	3		
wall	5		
Gassendi			
central peaks	6		
Grimaldi			
floor	1		

The present scale of brightness is decidedly better than nothing but it is inadequate for modern research. A suggested plan for creating a new and acceptable scale is outlined later in the book —in Chapter 16.

3 IDENTIFICATION OF LUNAR FEATURES

Features on the moon are located on maps, just as features are on the earth. The making of lunar maps and their use is called selenography. Almost all maps of the earth are drawn, but there are two types of lunar maps. All the famous old books of maps discussed in Chapter 1 are filled with drawn maps, to which were added detailed descriptions of the features. The latest of such books, by Wilkins and Moore, is the last in the distinguished line that began before the middle of the seventeenth century. Maps of this kind describe far more surface detail than can be found in any other source.

The other method of lunar mapmaking is by the use of photographs. These have the great advantage of absolute "accuracy," which cannot be achieved on a drawn map. But this "accuracy" is a tricky thing and considerable experience is needed in the proper interpretation of photographic maps. The tyro is likely to be much surprised and disheartened by the difficulty he experiences when he goes from map to telescope.

The principal reason for this failure of agreement between lunar maps and personal observation is the great variations in shadow which, as the moon's phase progresses, change the appearance of most lunar features by unbelievable amounts. Consider one of the finest lunar objec[t] the great ringed plain called Eratosthenes. Wh[en] the sun is rather low in the lunar sky, this pl[ain] (Plate 7–5) is highly conspicuous and exhib[its] a great deal of interesting detail. But at f[ull] moon, in the full blaze of reflected light (Pl[ate] 7–4) it can scarcely be seen! The cause is t[he] smallness of variation in the reflective qual[ity] of its rocky surface, so that shadows are requir[ed] if it is to be seen well enough to study. The r[ay] system of the neighboring crater Copernicus [is] an example of the opposite type of change. Und[er] a low sun Copernicus is little more spectacul[ar] than is Eratosthenes and its system of rays is d[if]ficult to observe. At the full moon, howev[er,] Copernicus and its rays stand out with explosi[ve] grandeur, and, along with the ray system [of] Tycho, rival the maria for pre-eminence. Ma[ny] hours of careful observing are necessary befo[re] the observer can avoid the pitfalls inherent [in] these apparent changes of the lunar surface. Ev[en] skilled selenographers have been known to ma[ke] embarrassing mistakes.

For the purposes of this book, any depressi[on] below the surrounding surface (except for tho[se] depressed areas that we designate as maria, v[al]leys, or rills) is defined as a crater, a term th[at] includes even many shallow depressions probably due only to the accidental shaping of an area. At our distance from the moon it is often impossible to distinguish these from true craters. The latter may be large or small and circular, oval, or polygonal. They may further be divided into two categories designated as "explosive" and "non-explosive." Those large depressions that for generations have been called mountain-walled plains comprise nearly all the important non-explosive craters. If any explosions occurred during the formation of these walled plains they were apparently merely secondary or "trigger-ing" events. (Such features are the subject of Chapter 11.) The explosive craters, on the other hand, were apparently caused by either volcanic activity or impact of some kind. (They are described in Chapter 12.) The explosive craters have been divided, classically, into ringed plains, craters, craterlets, and crater pits. There are many hybrids among the craters that exhibit both non-explosive and explosive factors in their formation; these should be classified according to which cause apparently was the more important. Alphonsus is an excellent example of such a hybrid.

The true craters can also be divided on the

Figure 6. Libration in longitude. The orbit of the moon is an ellipse with the mutual center of the mass of earth and moon at one focus. Because of the small mass of the moon, that point is only 3,000 miles from the center of the earth. We call the line that connects the center of the earth to the center of the moon, at any given instant, the *radius vector*. Newton's law of gravitation demands that this line sweep over equal areas each second of the month. Consequently when the line is short, the moon must move faster. Consider a month that happens to begin when the moon is at P, closest to the earth. Because of the short radius vectors the moon will move all the way to X in the first quarter of the month. The arrow represents some lunar feature that is directly toward the earth when the moon is at P. The moon turns on its axis only a quarter of the way around by the time it reaches X. Therefore we see the arrow displaced. By the time the moon reaches A, the arrow is again toward the earth. When the moon is at Y the arrow is seen from the earth as displaced to the other side from what it was at X. When the moon gets back to P at the end of the month, the arrow once more is toward us. The eccentricity of the moon's orbit is exaggerated in this diagram to make the displacement at X and Y, or libration, easily visible. Also we neglect the small angle between the planes of the moon's orbit and its equator. That neglected angle produces the libration in latitude (see Figure 7).

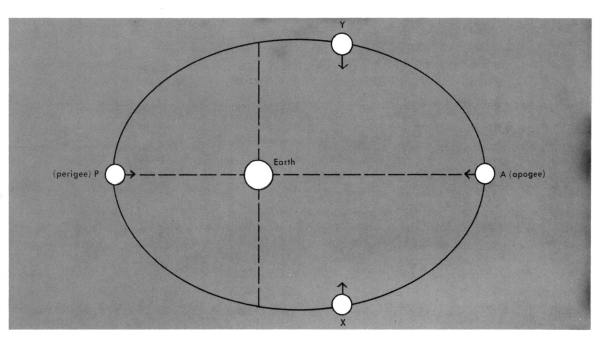

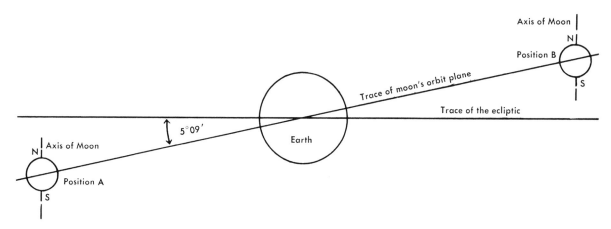

Figure 7. Libration in latitude. The observer is looking at the earth-moon system from a point out along the line in which the plane of the ecliptic intersects the plane of the moon's orbit. From this position the traces of both these planes are straight lines. The angle of 5°09′ between them is purposely exaggerated to make the libration more noticeable. The axis of the moon's rotation is almost perpendicular to the ecliptic, not to the plane of its orbit about the earth. N and S are the north and south poles of the moon. When the moon is in position A a man on the earth can see several degrees beyond the moon's north pole. When it is in position B, a half-month later, he can see approximately the same distance beyond the south pole.

basis of whether or not *external* impact was a factor in their origin. There is need for some single term to encompass loosely all craterlike formations that may be considered, in some context, as being non-impact. The name endocrater is suggested for this heterogeneous aggregation. Ectocrater could be used for all others, but the single adjective *impact* is widely used and can suffice.

Considerable difficulty in the identification of lunar features arises from the fact that the moon does not quite keep exactly the same face toward us. Certain features near the limb, or apparent edge, sometimes disappear "around the corner" as a result of this irregularity or tipping, and even when they are visible foreshortening can distort them very much. This tipping is known as *libration*, because of the resemblance of its back-and-forth motion to the up-and-down motion of a balance scale (libra is the Latin word for such a scale). The librations are responsible for irregularities of about 16° in the face which the moon turns toward the earth. There are four kinds of libration, two of which have a significant effect on the movements and appearance of the moon.

The first of these is *libration in longitude*. Despite the fact that the moon never can have

had any real oceans, the earth has produced tides on its surface. These tides have exerted a frictional drag on the moon, slowing its rotation just as the lunar tides are slowing the rotation of the earth today. The tidal effect of the earth on its relatively small satellite is so great that the moon's rotation has been slowed till it equals the period of its revolution about the earth and the moon is forced to keep one face toward the earth all the time. If its orbit around the earth were circular, the moon would circle the earth at a constant speed, completing each orbital revolution in constant phase with its rotation about its axis. But, the orbit is quite oval in shape. At the point where the orbit comes closest to the earth, the force of gravitation causes the moon to increase its speed. A half-month later, when it is farthest from the earth, the lessened force of the earth's gravity has caused the moon to slow down. During this time of decreased orbital speed the moon continues to rotate about its axis at a constant speed, and we, therefore, see somewhat more of its surface at one edge than we would have if both speeds had continued constant. During the time of increased orbital speed, the rotation of the moon falls behind, as it were, and we can see the same extra amount of its surface at the other

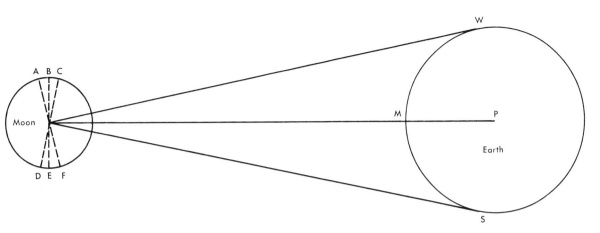

Figure 8. Diurnal libration. In this diagram the observer is looking down from a position above the earth's north pole. He sees the earth turning counterclockwise about its pole in a day and the moon revolving in the same direction about the earth in a month. When the daily spinning of the earth carries a person to the point W, the moon rises for him. When it has taken him to M the moon is on his meridian and when he reaches S the moon is setting. When the moon is rising he can see all the way around to A, but at the opposite limb cannot see beyond F. When the moon is on his meridian his vision is limited to B and E. When the moon is setting the boundaries of the half visible to him are C and D. In other words, he can see alternately a degree beyond the mean position of each limb.

limb. As a result of all this, at one time we can see nearly eight degrees "around" the eastern limb, and a half-month later an equal distance at the western limb instead becomes visible. Large librations of opposite sign on Plates 4–3 and 4–9 change the position of Mare Symthii from nearside to farside.

The second important libration is the *libration in latitude*, which results from a similar tipping in the north and south direction. The plane of the moon's equator of rotation is not in the plane of its orbit around the earth but makes an angle of a little less than seven degrees with it. As a result, at one time of the month we can see about seven degrees past the north pole and a half-month later seven degrees past the south pole. On Plate 4–3 the moon is tipped by almost the maximum amount to the south. On Plates 4–9 and 5–2 it is tipped as much to the north.

Diurnal libration is rather small and is due only to the diameter of the earth. The accompanying diagram, which of course is not drawn to scale, may assist in understanding it. The diameter of the moon, as seen from the earth, is half a degree, while that of the earth, as seen from the moon, is two degrees. The average distance between their centers is thirty times the

diameter of the earth. As a result, the moon seems to tip slightly between its rising and setting.

Finally, the pole of the moon's rotation is not quite 90° from the plane in which it and the earth move about the sun, missing by about 1½ degrees. This produces a fourth tiny libration, which is of interest only to certain astronomical specialists.

The longitude and latitude of lunar features are measured in the same manner as for terrestrial features. The coordinate origin, or reference point, is the point that would be the center of the moon's visible disk if all librations were simultaneously at their zero values. (On the earth the comparable reference point is the intersection of the Greenwich meridian with the terrestrial equator.) Longitude is measured 360° around the equator from this point. That is, an observer standing on the moon and facing south will have increasing longitude toward his left. Still another way of saying it is that for an observer at the north pole, longitude increases counterclockwise. As on the earth, latitude is measured along meridians and reaches 90° at the poles. We call these coordinates selenographic longitude and latitude.

The librations cause this reference point to

oscillate irregularly around the center of the actual visible disk. As a result the height of the sun above the horizon for any point on the moon's surface is different this month from what it will be at exactly the same phase next month. For example, the sun will not rise on Copernicus at quite the same phase in successive months. The difference is great enough to bother any systematic observer. *The American Ephemeris* * tabulates for the beginning of each day of the year a quantity called the "colongitude of the sun." This quantity is equal to the negative of the longitude of any feature on the moon's equator for which the sun is rising at the instant. It gives a good approximation to the time of sunrise for a wide belt on each side of the equator. It is tabulated for each feature of the list at the end of this chapter, except for those which are too close to the poles for it to have much value. A more complete explanation is in the appendix on pages 199–200.

This chapter includes, as Plates 3–1 and 3–2, a very comprehensive, photographic map of the moon. It is composed of two half moon photographs from the famous series made by J. H. Moore and J. F. Chappell, using the 36-inch refractor of the Lick Observatory at Mount Hamilton. Their procedure is described on page 197. (A slight difference in the librations causes only minor difficulty in use of the two chosen pictures as a map, and that little is found only in high latitudes.)

These photographs have been superposed on an arbitrary grid with x and y coordinate numbers running to the right and upward from the lower left corner. This coordinate system has no negative numbers and is convenient for identifying features. (For almost any other study of the lunar system the scale adopted by the International Astronomical Union in 1932 should be used.) As is almost universally the practice today among astronomers, south is mounted at the top and the direction which has habitually been called east at the right hand side. This causes the reader to see the moon on his map in the

same orientation as he sees it in an astronomical telescope. The astronomical convention has been to define east on the moon as toward the limb that we see to the east in our own sky. This arbitrary convention was convenient for the early astronomers but it has caused the sun to rise in the west on the moon despite the fact that the moon rotates on its axis in the same direction that the earth does. Until this decade that has been unimportant.

The International Astronomical Union adopted at its triennial meeting in August 1961 the following resolution:

"Resolution Number 1

"For compiling new maps of the Moon, the following conventions are recommended:

"Astronomical maps for purposes of telescopic observations are oriented according to the astronomical practice, the South being up. To remove confusion the terms East and West are deleted.

"Astronomical maps for direct exploration purposes, are printed in agreement with ordinary terrestrial mapping, North being up, East at right and West at left.

"Altitudes and distances are given in the Metric System."

East and west on the topographical maps are the exact opposite of the older astronomical books. Almost every book to date has followed the former custom. In this text, a man standing on the moon and facing south will have sunrise on his left and sunset on his right. For the next few years the changing convention will be troublesome but under twentieth-century space plans it is necessary.

An alphabetic index of 243 features of the moon will be found at the end of this chapter. The x and y coordinates are given, and also the longitude, in order that the user may have no difficulty in determining the approximate times of sunrise, except for features far from the equator. With the exception of Lyot and a few descriptive names enclosed in quotation marks, all the names are from the list approved by the International Astronomical Union at its 1932 and later sessions and used in the I.A.U. catalogue and maps under the title of "Named Lunar Formations." There have been inconsistencies in the past, especially in the spelling of Latinized

---

* *The American Ephemeris and Nautical Almanac* which can be consulted in any good library or may be purchased each year from the U.S. Superintendent of Documents, Washington, D.C. The beginning of the day is the instant when Greenwich Mean Time, now called Universal Time, is 00ʰ 00ᵐ 00ˢ.

names. The official spellings are used here even in those cases where a different one might be preferable. It is better to have worldwide acceptance than to attempt to improve certain usages.*

There is a very easy way to use this map to identify any feature named in the index. Lay a piece of heavy bond paper along the x scale at the top of the map and carefully mark on it the grid coordinates printed there. To identify a feature seen on the map, lay the paper scale horizontally across the center of the object and read the x coordinate from the scale where it crosses the feature and the y coordinate from the scales at the sides of the map and then look for those coordinates in the index. To find an object for which the coordinates are given, place the scale horizontally at the given y value and then look for the object where the scale shows the given x coordinate.

For large objects such as the maria and some of the mountain-walled plains, the listed coordinates apply to a point near the center of the object. Care has been exercised to avoid designating large features by coordinates that are close to some prominent smaller object within the feature. Otherwise a beginner who did not know, for example, whether Mare Imbrium is large or small might become confused.

A photographic map is at a disadvantage in showing features that lie so close to the limb that they become unobservable when the librations are unfavorable, but are rather easy to observe when the librations are favorable. No ordinary type of map could possibly show such a feature (the walled plain Bailly is an example) under both these conditions; a globe would be required. In such cases a drawn map may have an advantage over the photographic kind. Usually it is made for the mean libration position, and it can even be distorted slightly to show a bit of what lies around the limb. But drawn maps have shortcomings. First, almost all of them are inaccurate in detail. Moreover, the fact that most of the features are drawn merely as lines, without any of the subtle differences in shading, is disadvantageous.

* The maps following in this chapter are in the astronomical convention, the South being up. The maps in Chapter 19 are in the astronautical convention, the North being up, like maps of the earth, for direct exploration.

The nearest approach to a perfect map (other than a globe) would require hundreds of pages of photographic prints. It would be produced for a dozen colongitudes and various combinations of librations. Some two dozen sheets would be required to cover the lunar surface for each situation; the pages would be very large and there would be a total of some 1,000 photographic prints. Except for use in the most specialized lunar research, the advantages would not be worth the cost and the bulk. A practical approximation of such an ideal map is the large lunar atlas made by Dr. Gerard P. Kuiper and published in 1960 by the University of Chicago. The supplementary atlas made by D. W. G. Arthur and E. A. Whitaker from some of Kuiper's sheets is the most useful yet prepared for the student whose research requires accurate quantitative data.

The almost 250 features listed below are not necessarily the most important ones. There would be considerable disagreement with respect to any such selection, and any individual would probably change his own choices from year to year. Primarily the listed features are the ones that are mentioned in this book, with some additional objects included, particularly in thinly identified areas. The International Astronomical Union has made nearly 700 lunar feature designations official, with several thousand additional sub-designations.

To prepare a detailed map which will be convenient and yet will satisfy most of the requirements of both the amateur and the professional astronomer, the moon has been divided into 30 overlapping sections. The four "corner" ones have been mounted as one—Plate 3-3. The other 26 comprise the large photographs, Plates 3-4 to 3-29. Each of the features included in the following list has been labeled on one or more of these maps. For the largest features, the name tab is either repeated in various locations or is placed near the center of the object, for example, Oceanus Procellarum on Plates 3-21, 3-24, and 3-27. To make these sections, the Lick Observatory pictures mounted on the grid as Plates 3-1 and 3-2 were much enlarged and then cut. Use of the same two plates for the coordinate maps and the enlarged sections makes identification very easy.

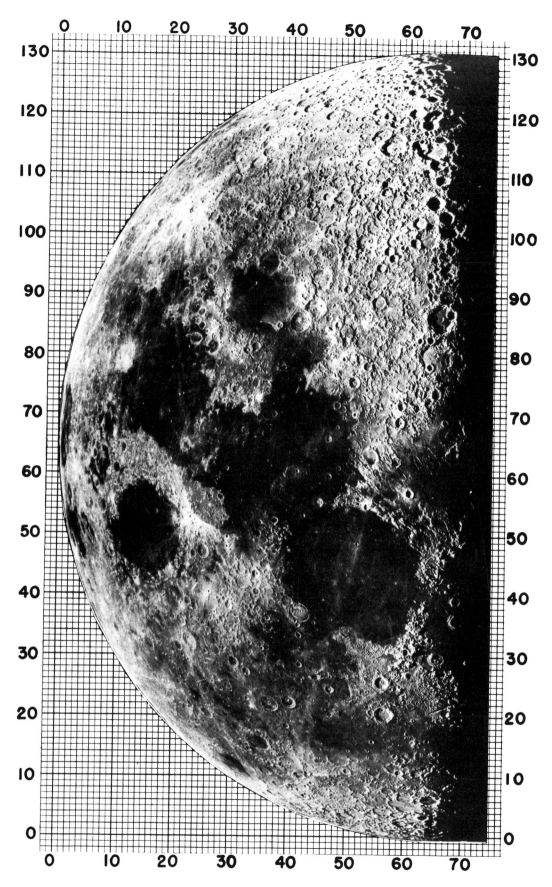

Plate 3–1. Lick Observatory, 1938 May 7^d 04^h 47^m UT, phase 6.97 days, colongitude 0°. This and Plate 3–2 form a general map. See pages 20–21 for explanation of grid and Table 4 for the grid coordinates and the colongitudes of the sun at time of sunrise for 243 lunar features. Plates 3–4 to 3–16 and part of Plate 3–3 are enlarged, overlapping sections of this plate.

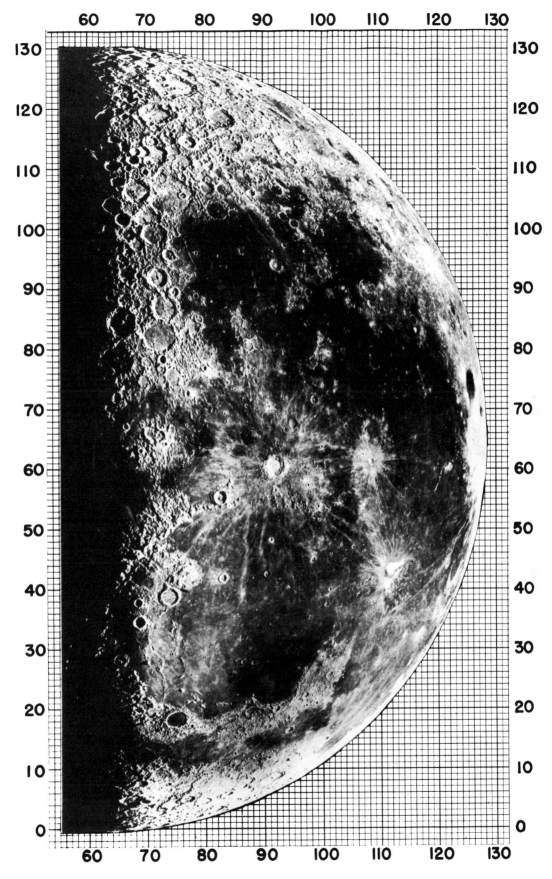

Plate 3–2. Lick Observatory, 1937 October 26^d 13^h 41^m UT, phase 22.07 days, colongitude 174°. Plates 3–17 to 3–29 are enlarged sections of this plate.

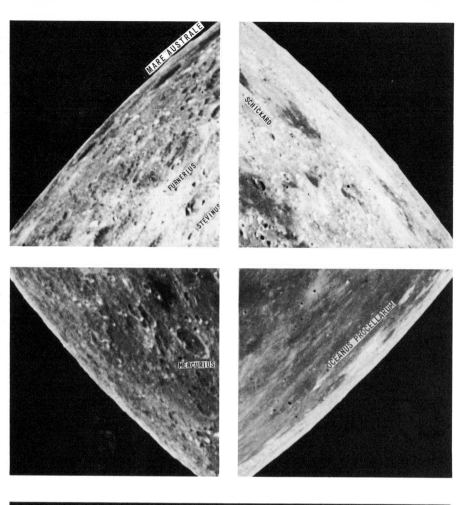

Plate 3–3. Left to right: upper left, upper right, lower left, and lower right sections of Plates 3–1 and 3–2.

Plate 3–4. Notable in this section are the Furnerius-Stevinus ray system, the Rheita depression, Mare Australe, Janssen, and long scarps which have been partly destroyed by more recent activity. See Plate 3–1 for technical details.

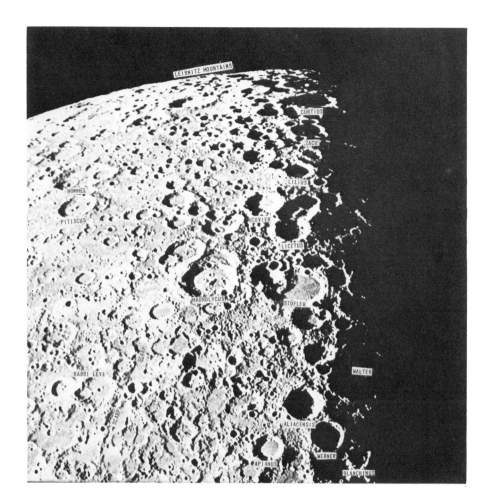

Plate 3–5. In the southern (top) part of this picture is a vast, mostly right and left depression which extends almost across the section. The boundaries are so low, and it has been so nearly destroyed by later changes that it is observable only under a low Sun. See Plate 3–1 for technical details.

Plate 3–6. Mare Foecunditatis with Stevinus, Petavius, Wendelinus, and Langrenus in line. Lower right Messier and W. H. Pickering with the famous double ray. See Plate 3–1 for technical details.

Plate 3–7. Theophilus wi[th] Mare Nectaris and Frac[as]torius. The Altai Scarp is [at] the upper right. A great sca[rp] runs southward from Gute[n]berg to the left-hand shore [of] Nectaris. See Plate 3–1 f[or] technical details.

Plate 3–8. Area to the rig[ht] of the Altai Scarp. Notice t[he] very long irregular sca[rp] southward from Albategni[us] to southern edge of the se[c]tion. It is the right bounda[ry] of a depression which [is] bounded on the left by [an] arc of six great ringed plain[s.] See Plate 3–1 for technic[al] details.

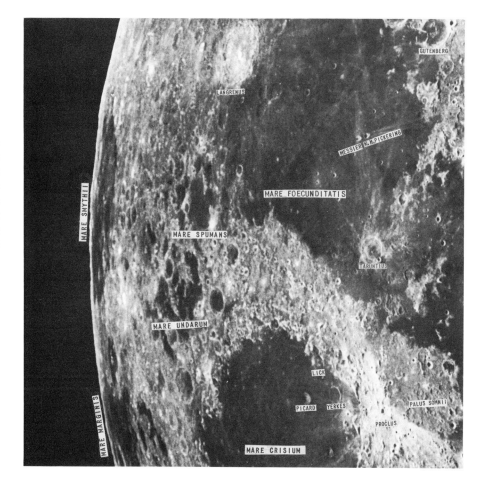

late 3–9. The great group f "Western Maria." Maria pumans and Undarum are omposed mostly of not enrely merged small sunken reas. Mare Smythii is almost e size of Mare Crisium but ffers from extreme right-left reshortening. See Plate 3–1 r technical details.

late 3–10. The section consts, primarily, of Mare Tranuillitatis. See Plate 3–1 for echnical details.

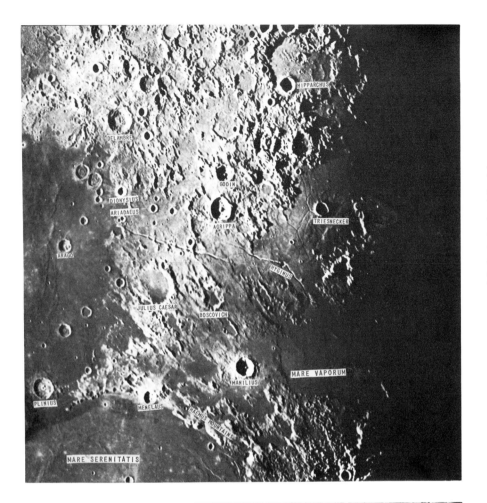

Plate 3–11. This shows th[e] greatest of rill systems, th[e] triple system of Triesnecke[r] at the terminator, Hyginu[s] and Ariadaeus. At the upp[er] edge is Hipparchus, a ve[ry] old, almost obliterated walle[d] plain. Note the "finger" [of] smoother surface down an[d] left from it to the Ariadae[us] rill. See Plate 3–1 for techr[i]cal details.

Plate 3–12. Note the "ghost[s]" on the northern floor of Ma[re] Crisium. Note that there a[re] no rays to the upper rig[ht] from Proclus through Pal[us] Somnii. The libration is u[n]usually favorable to Ma[re] Marginis. See Plate 3–1 f[or] technical details.

Plate 3–13. Proclus is in the upper left. Posidonius is on the left hand of Mare Serenitatis. See Plate 3–1 for technical details.

Plate 3–14. Mare Serenitatis and the strait which connects it to Mare Imbrium. Notice the great ridge in the left of Serenitatis. Linné is on the right part of the floor. Aristillus and Autolycus are on the terminator. See Plate 3–1 for technical details.

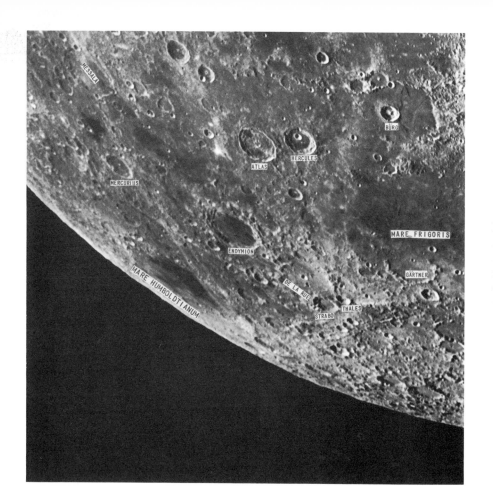

Plate 3–15. Notice Atlas and Hercules. The variable black spot on the floor of Atlas makes it a favorite of observers. See Plate 3–1 for technical details.

Plate 3–16. The Alpine Valley is at the terminator. Aristoteles and Eudoxus are above the central part of the section. Mare Frigoris crosses the center just below Aristoteles. See Plate 3–1 for technical details.

Plate 3–17. Clavius to Deslandres. Notice that Clavius (the second largest of all walled plains) lies in the right part of a still older but much larger depression, now barely observable and only seen at all when near the terminator. Tycho is rather inconspicuous under a low sun. See Plate 3–2 for technical details.

Plate 3–18. Area right of Clavius and Tycho. This section contains the much elongated walled plain, Schiller. Near the limb is the tremendous walled plain Schickard, actually a small mare. Just above it is strange Wargentin. Unfortunately librations hid Bailly, which was just beyond the limb when the picture was made. It is larger than Clavius and should be classed as a mare. See Plate 3–2 for technical details.

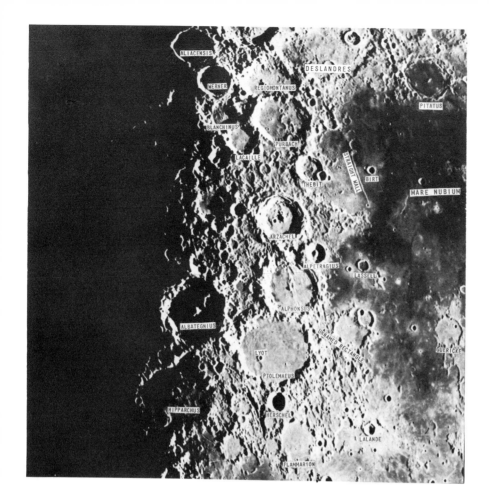

Plate 3–19. Hipparchus, Albategnius, Ptolemaeus, Alphonsus, Arzachel, Pitatus, Straight Wall, and Mare Nubium. See Plate 3–2 for technical details.

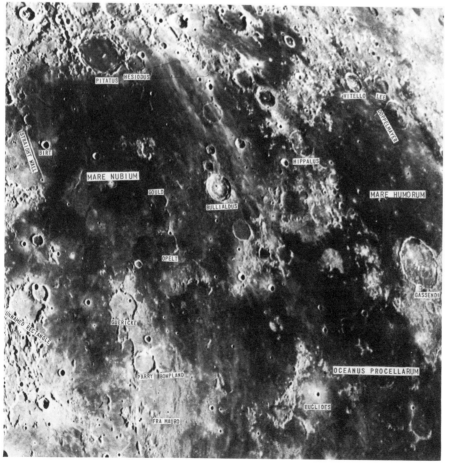

Plate 3–20. The Maria Nubium and Humorum. Note the great double ray from Tycho crossing Nubium. Note Gassendi at northern shore of Humorum. "Ghosts" are conspicuous on the floor of Nubium, which received its name from them. See Plate 3–2 for technical details.

Plate 3–21. The extremely bright ringed plain Byrgius is right of Mare Humorum and near the limb. Notice the great "ghost" Letronne directly north of Gassendi. See Plate 3–2 for technical details.

ROOK MOUNTAINS

VITELLO
LEE
DOPPELMAYER

MARE HUMORUM

BYRGIUS

CORDILLERA MOUNTAINS

MERSENIUS

GASSENDI

LETRONNE

WICHMANN

OCEANUS PROCELLARUM

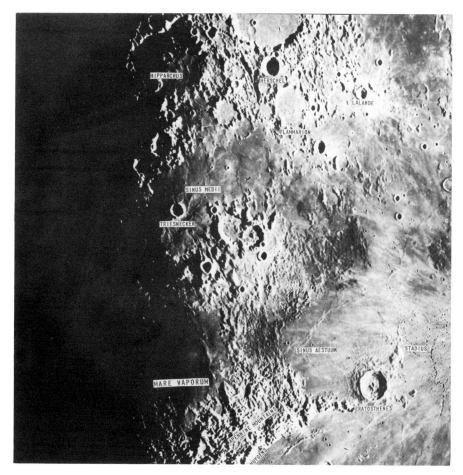

Plate 3–22. Central part of the lunar disk. Note the rough area north of Sinus Medii. It is almost a large island but is connected by an isthmus to the Apennines. Note the two rough black areas on its right. They are almost unique. The Apennines have many small craters on them. See Plate 3–2 for technical details.

HIPPARCHUS
HERSCHEL
LALANDE
FLAMMARION

SINUS MEDII

TRIESNECKER

SINUS AESTUUM
STADIUS

MARE VAPORUM

APENNINE MOUNTAINS
HUYGENS
ERATOSTHENES

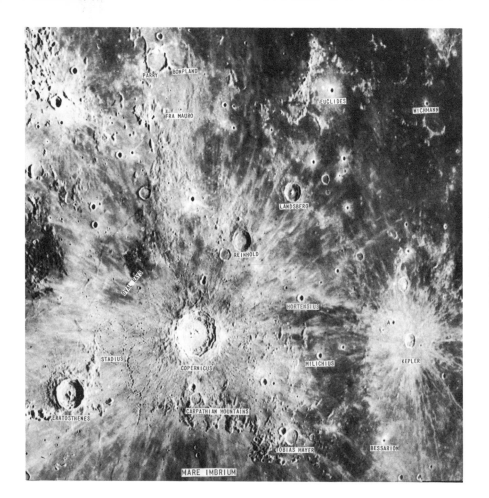

Plate 3–23. The Copernicu[s] area. This region is the sub[-] ject of Chapter 7. The ra[y] system is one of the greate[st] features of the moon. Se[e] Plate 3–2 for technical d[e-] tails.

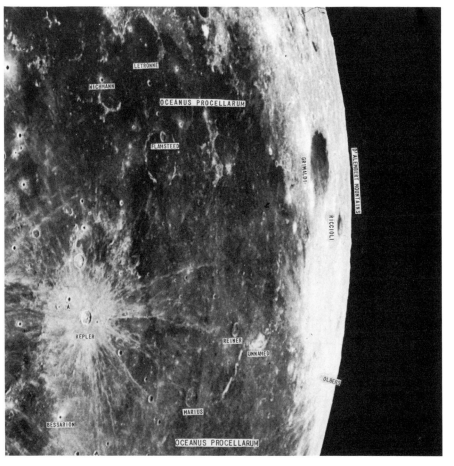

Plate 3–24. Kepler and i[ts] ray system, right to the lim[b.] Near the limb is Grimald[i,] which should be classed as [a] small sea. See Plate 3–2 f[or] technical details.

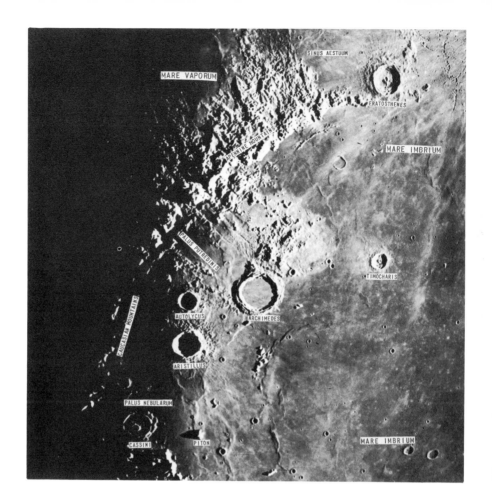

Plate 3–25. Apennines and northward along the left-hand part of Mare Imbrium. Note Archimedes and its rill system, the narrow radial valleys leading from Aristillus, and the ridges in western Imbrium. See Plate 3–2 for technical details.

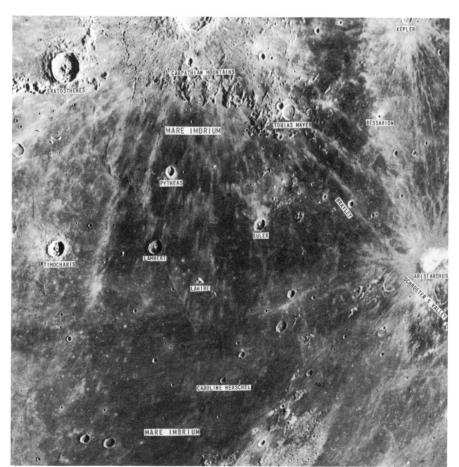

Plate 3–26. Mare Imbrium. Note the complex structure of rays from Copernicus. See Plate 3–2 for technical details.

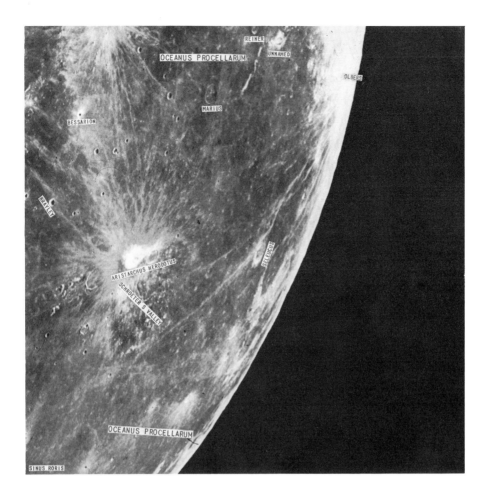

Plate 3–27. Aristarchus a[n] its ray system, interconne[c] ing with systems of Coper[ni] cus, Kepler, Olbers (at e[x] treme right), and Seleuc[u] Aristarchus is brightest fe[a] ture on the moon. Herodot[us] is just right of it. Schroete[r] Valley is north of the tw[o] Note the disturbed ground [to] their north. See Plate 3–2 f[or] technical details.

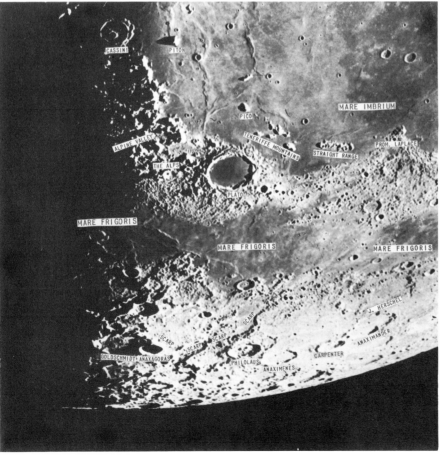

Plate 3–28. Plato with Ma[re] Imbrium to the south a[nd] Mare Frigoris to the nort[h] Note the ridges in the l[eft] part of Imbrium. See Pla[te] 3–2 for technical details.

Plate 3-29. At Sinus Iridum (the Bay of Rainbows). This is the remnant of a large walled plain. The partial destruction was similar to that at the Straight Wall. The right end of Mare Frigoris is just below it. See Plate 3-2 for technical details.

Table 4

Name of feature	x	y	Selenographic colongitude at sunrise	Name of feature	x	y	Selenographic colongitude at sunrise
Abenezra	57	96	349°	Aristoteles	56	20	343°
Abulfeda	56	88	346	Arzachel	73	92	3
Agrippa	58	69	350	Atlas	57	22	316
Albategnius	66	84	356	Autolycus	69	38	358
Aliacensis	65	105	355	Azophi	56	97	348
Almanon	55	91	345	Bailly	85	126	65
Alphonsus	74	87	4	Ball	77	109	9
Alpine Valley	67	21	357	Barrow	64	6	355
Alps, mountains	67	20	0	Beaumont	39	92	352
Altai Scarp	47	98	340	Bessarion	107	54	37
Anaxagoras	71	6	11	Bessel	52	48	342
Anaximander	87	8	40	Biot	21	95	310
Anaximines	79	4	40	Birt	79	96	9
Apennine Mountains	72	50	5	Blancanus	77	126	21
Apianus	61	101	352	Blanchinus	67	99	358
Arago	47	66	358	Bond, W. C.	66	10	357
Archimedes	75	39	4	Bonpland	89	81	17
Ariadaeus	50	66	342	Boscovich	58	62	349
Aristarchus	112	43	47	Brayley	106	47	37
Aristillus	69	34	359	Bullialdus	92	94	22

Name of feature	x	y	Selenographic colongitude at sunrise	Name of feature	x	y	Selenographic colongitude at sunrise
Bürg	47	24	332°	Hercules	40	22	322°
Byrgius	118	94	65	Herodotus	113	43	49
Carpathian Mountains	95	55	20	Herschel	73	78	3
Carpenter	84	6	46	Herschel, Caroline	97	33	31
Cassini	65	25	356	Herschel, John	86	10	40
"Cassini's Bright Spot"	73	106	4	Hesiodus	85	103	16
Catharina	45	92	337	Hind	62	82	352
Caucasian Mountains	62	32	350	Hind C	61	82	351
Censorinus	35	73	328	Hippalus	98	97	30
Clavius	75	124	15	Hipparchus	65	78	354
Cleomedes	20	40	304	Hommel	50	121	326
Conon	68	49	358	Hortensius	100	64	28
Copernicus	93	60	20	Huggins	70	113	3
Cordilleras, mountains	121	93		Huygens, Mount	73	49	4
Curtius	64	127	356	Hyginus	64	65	354
Cuvier	61	121	351	Jansen	40	56	351
Cyrillus	44	87	335	Janssen	38	114	320
D'Alembert Mountains	127	76	90	Julius Caesar	54	63	345
Delambre	51	75	342	Kepler	108	61	38
De La Rue	41	12	310	Kepler A	107	62	36
Deslandres	74	106	5	Lacaille	69	97	358
Dionysius	51	70	343	Lahire	94	40	25
Doerfel Mountains	89	125		Lalande	80	77	9
Doppelmayer	106	101	41	Lambert	91	42	21
Endymion	35	15	306	Landsberg	99	71	26
Eratosthenes	83	55	12	Langrenus	11	79	300
Euclides	101	80	29	Laplace, Prom.	88	22	26
Eudoxus	56	25	344	Lassell	79	88	8
Euler	98	45	28	Lee	104	103	40
Fabricius	35	113	319	Leibnitz Mountains	55	130	
Flammarion	75	75	4	Le Monnier	39	42	330
Flamsteed	113	75	44	Letronne	112	83	42
Fracastorius	36	95	327	Lexell	73	108	5
Fra Mauro	90	79	17	Licetus	63	119	354
Furnerius	20	105	300	Lick	18	56	308
Gärtner	49	13	326	Lilius	64	123	354
Gassendi	108	90	40	Linné	58	42	348
Geber	55	94	346	Longomontanus	84	118	22
Geminus	22	35	304	Lyot	72	81	1
Godin	58	71	350	Macrobius	25	47	314
Goldschmidt	67	5	3	Mädler	38	85	330
Gould	88	92	17	Maginus	72	119	7
Grimaldi	126	74	67	Manilius	60	57	351
Gruemberger	70	127	12	Mare Australe	25	115	275
Guericke	86	84	14	Mare Crisium	15	50	300
Gutenberg	26	81	318	Mare Foecunditatis	18	75	305
Hadley, Mount	65	41	355	Mare Frigoris	45–100	14	
Haemus Mountains	60	50	345	Mare Humboldtianum	31	12	285
Harpalus	94	16	44	Mare Humorum	105	96	30
Hell	77	105	8	Mare Imbrium	90	35	20
Heraclides, Prom.	96	26	35	Mare Marginis	3	50	280

Name of feature	x	y	Selenographic colongitude at sunrise	Name of feature	x	y	Selenographic colongitude at sunrise
Mare Nectaris	35	90	330°	Pytheas	91	48	21°
Mare Nubium	83	94	15	Rabbi Levi	48	109	336
Mare Serenitatis	52	42	340	Ramsden	97	105	31
Mare Smythii	1	68	280	Regiomontanus	71	102	2
Mare Spumans	8	67	295	Reiner	116	65	55
Mare Tranquillitatis	40	62	330	Reinhold	95	68	23
Mare Undarum	8	61	292	Rheita	30	109	314
Mare Vaporum	65	58	355	Rheita Valley	31	112	314
Marius	114	56	50	Riccioli	127	70	75
Maurolycus	57	115	345	Rook Mountains	110	108	90
Menelaus	53	55	344	Rosenberger	47	122	318
Mercurius	26	21	295	Rosse	34	92	326
Mersenius	113	93	48	Rutherford	72	125	13
Messala	23	32	302	Scheiner	80	124	27
Messier	21	73	313	Schickard	104	113	55
Metius	33	112	317	Schiller	93	119	39
Meton	61	5	338	Schroeter's Valley	113	42	49
Milichius	102	60	30	Seleucus	122	44	66
Miller	68	111	0	Sinus Aestuum	80	58	10
Moretus	67	128	7	Sinus Iridum	93	22	32
Nasireddin	68	113	1	"Sinus Iridum Highlands"	90	20	35
Newton	70	130	18	Sinus Medii	72	70	0
Oceanus Procellarum	120	35–85	45	Sinus Roris	104	23	50
Olbers	127	59	76	Snellius	21	101	304
Opelt	88	90	17	Stadius	86	60	13
Orontius	72	112	4	Stevinus	23	104	306
"Oval Rays"	86	64	14	Stöfler	62	115	354
Palus Nebularum	70	31	0	Strabo	42	10	307
Palus Putredinis	70	42	0	Straight Range	83	21	21
Palus Somnii	25	54	318	Straight Wall	78	96	8
Parry	88	80	16	Taruntius	22	64	314
Petavius	16	96	300	Taurus Mountains	30	35	320
Philolaus	77	5	30	Teneriffe Range	79	21	15
Phocylides	98	119	56	Thales	44	10	310
Picard	17	54	306	Thebit	75	90	5
Piccolomini	39	103	328	Theophilus	42	85	333
Pickering, W. H.	22	73	314	Timocharis	84	42	13
Pico, Mount	76	23	9	Tobias Mayer	99	53	29
Pitatus	83	103	13	Triesnecker	66	69	356
Pitiscus	49	117	330	Tycho	78	114	12
Piton, Mount	71	28	2	Vendelinus	12	87	300
Plato	76	18	9	Vitello	102	102	37
Playfair	61	99	302	Vlacq	48	120	323
Plinius	45	55	336	Walter	70	106	359
Polybius	43	97	335	Wargentin	103	115	60
Posidonius	42	36	331	Werner	67	102	357
Proclus	23	52	314	Wichmann	108	79	37
Ptolemaeus	74	82	3	Wilhelm I	85	114	20
Purbach	72	99	3	Wrottesley	18	96	304
Pyrenees Mountains	28	87	320	Yerkes	21	53	310
Pythagoras	90	8	60	Zach	64	125	355

4 THE WAXING MOON

Almost everyone is awed by his first look at the moon through a telescope. Observing it is fascinating for a while. But interest palls soon unless one begins to study special features in sufficient detail to become quite familiar with them. This chapter and the two that follow consider the moon as a whole. They contain only brief descriptions of some of the most important features. Detailed accounts of most of the objects mentioned here and of certain others, with explanations of their possible origins, are given in later chapters and illustrated by numerous enlarged photographs. It is hoped that these three general survey chapters will induce some of the thousands who own adequate telescopes to use their instruments systematically and to learn at first hand of some of the wonders that are visible. For the more casual reader photographs will suffice quite well.

When the thin crescent moon can first be observed it appears as on Plate 4–1. The photograph exhibits the moon at phase 2.66 days—that is, 2.66 days after it was most closely in the direct earth-sun line. At this phase the moon is still so nearly in line with the sun that by the time the sky is dark enough for a photograph to be made the moon is very low in the west. The

mountings of most of the great telescopes will n∘ permit observations so close to the horizo⸱ Moreover, at low altitudes the atmosphere is usᵤ ally too unsteady to permit crisp photograph images. The amateur, visual observer will als have difficulty with his portable telescope at th phase, for the air is "boiling" too much in his fiel of view to allow him to see fine detail. On th March evening when this photograph was mad⸱ conditions were unusually good. It should ⸀ noted that the sunrise terminator has bare reached the left shore of Mare Crisium.

By the next evening (Plate 4–2) a very di⸱ ferent landscape may be observed. Not only h⸱ the terminator moved far into areas entirely da⸱ the night before, but some of the features appe⸱ to be entirely different in form from what the⸱ were on the first evening. During the month th observer will discover that an interval of on⸱ a few days causes the appearance of such crate⸱ as Furnerius, Tycho, and Copernicus to chang so much that they are no longer recognizabl⸱ The changing phases of the moon that cause th appear in Figure 9 as seen by the naked eye ar⸱ in Figure 10 as viewed with a telescope. But if tl directions given in Chapter 3 are followed, m⸱ objects can be identified without difficulty. C

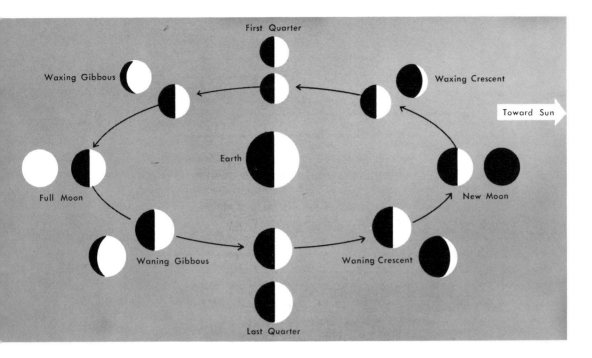

Figure 9. Phases of the moon. As the moon travels around the earth, one-half is in sunlight and one-half in darkness. However, varying amounts of the sunlit hemisphere are turned toward the earth, creating the phases seen from earth, as moon is viewed without a telescope.

of these views the usual astronomical conven-
of placing north at the bottom is followed.
sun rises on each equatorial object when the
ngitude at the moment equals the negative of
longitude. See table 4 in Chapter 3. The rule
ds quite well throughout a wide belt on each
of the equator.

During the phases of four to six days the
on displays a wide variety of features for ex-
ination in its early morning sunshine. One
y conspicuous example is Mare Crisium, the
of Crises. During this interval, as shown by
tes 4–2—4–7, its floor exhibits much lacy de-
. Part of this, especially in the northern por-
n, is composed of "ghosts"—craters that ap-

parently predated the mare and sank with the surrounding area during its formation. It seems probable that they were almost completely engulfed by the sea of molten rock, but not quite completely enough to obliterate all traces of their existence. When the sun is practically at the horizon it is actually possible to observe short shadows from a couple of the sunken walls. The most conspicuous of the ghosts still rise very slightly above the surface of the "sea." If the maria had been the results, primarily, of great explosions we would not expect to find these traces from the past.

The observer should also look (Plates 4–2 and 10–9) for the "continental shelf" that surrounds

Figure 10. Phases of the moon from waxing crescent (right) to full moon (center) and waning crescent (left), as photographed by the 100-inch telescope of the Mount Wilson and Palomar Observatories.

Plate 4–1. Griffith Obse[r]
tory, 1959 March 12^d 02^h
UT, phase 2.66 days, colo[ngi]
tude 298°. Paul E. Ro[c]
with 12-inch Zeiss refra[c]
Kodak Royal Pan Film, C
3385 filter (yellow). Pri[nci]
pally the left shore of N[are]
Foecunditatis.

the central part of Mare Crisium. It produces a sudden deepening of the floor, exactly as do the continental shelves of our terrestrial oceans. Ridges are common on mare floors but the boundary of this central area is not a ridge. Under a low morning sun the left hand boundary is darker than the neighboring floor, while the

right hand boundary is lighter. This is rever[sed] during the time before sunset. The boundary[is] a true scarp, and it provides strong evidence t[hat] not all of the sinking occurred at one time, [but] that a final subsidence came after the rock h[ad] lost much of its fluidity.

All around the shoreline of Mare Crisi[um]

an be seen the remains of large, old, craterlike features whose seaward walls apparently collapsed in the formation of the sea. Such remnants re common around most of the maria. In some ases they have become mere bays. In other cases heir seaward walls may still be observed under roper illumination. They present part of the vidence of the process by which the maria were rmed.

In the first days during which the crescent 1oon can be observed in the evening sky, the erminator moves to the right across Mare Foe-unditatis, which is south of Mare Crisium. In late 4–2 it is just entering that area. On Plate –2 the ridges in the left hand part of that mare how nicely and the terminator has just passed 1e craters Messier and W. H. Pickering. Shadow ffects cause the apparent sizes and shapes of

these two features to change during the month. It is one of the most common characteristics of human beings to prefer an exciting explanation of data to the prosaic one that is usually true. This is illustrated beautifully by some of the spectacular things which have been written about these two conspicuous little craters. Chapter 15 includes a series of enlarged pictures of them (Plate 15–8) and considers what actually takes place between sunrise and sunset. The double ray extending eastward from the pair should be followed carefully. Special note should be made of the fact that the southern component of the ray is a series of bright spots. The two rays end at a crater similar in size to the primary pair but less conspicuous because of its brighter background.

Sunrise on Mare Nectaris should also be ob-

late 4–2. Lick Observatory, 937 September 9ᵈ 02ʰ 45ᵐ T, phase 4.16 days, colongi-de 316°. Terminator at heita Valley, Mare Foecun-tatis, and Proclus.

served at this time of the month. In Plate 4–3 the terminator lies across it and reveals one of the most beautiful scenes on the moon. Two scarps show exquisitely. One runs southward from Gutenberg and eventually forms part of the shoreline on the left of Nectaris. The other, nearly parallel to it, is a continental shelf within the mare. The crater Rosse can be seen on this picture almost exactly at the terminator. On Plate 4–4 the terminator is beyond Nectaris and just outlines the crest of the Altai Scarp, far to the right. This scarp is hundreds of miles long and forms an arc centered on Mare Nectaris. There is little doubt that when Nectaris was formed the sinking extended to the right and upward far beyond its shoreline.

Plate 4–5, a day later, shows the whole area beyond Nectaris and reveals that it is less rugg left of the Altai Scarp than it is farther to th right. Plate 6–3, where the arc shows dark und a setting sun, gives adequate proof that this actually a great scarp and not a ridge, as som times has been thought.

Furnerius and Stevinus are visible in th upper left portion of Plate 4–2 and at this m ment there is nothing to attract attention to the On Plate 4–4 a definite brightening has occurre But by the time the colongitude of Plate 4–5 h been reached, a craterlet on the outer right har slope of each of them has brightened so co spicuously that the eye is caught immediatel A single day later this brightening has increase noticeably, and the great, vivid streaks of a con posite ray system can be seen. From then unt

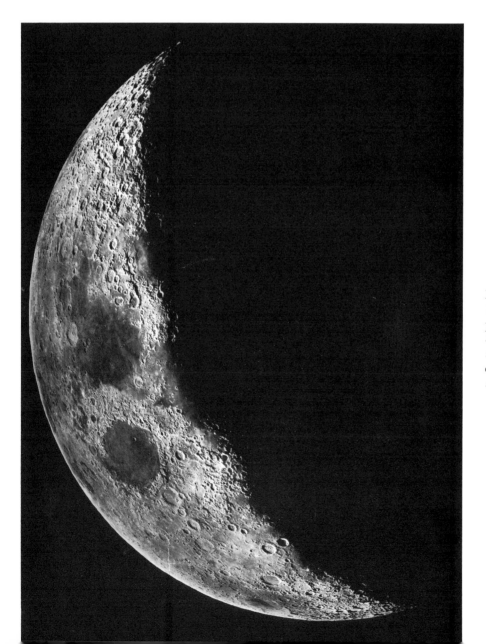

Plate 4–3. Lick Observator 1938 June 3ᵈ 04ʰ 02ᵐ U phase 4.59 days, colongitu 329°. Terminator at Jansse at Mare Nectaris, to the rig of Mare Tranquillitatis, an at Hercules.

ear sunset, long after full moon, their brightness
npresses even a casual observer. Shortly before
ll moon (Plate 4–11) this ray system can be
bserved to tie in with that of Tycho to form
ne tremendous oversystem.

Just before the half moon phase (Plate 4–5)
ie terminator crosses Mare Serenitatis in the
orthern part of the moon. This mare has a spec-
acular dark edging around its shoreline and is
orthy of much study. A great north and south

ridge in the left hand portion of the floor is a
continuation of the system that begins near Mare
Nectaris and proceeds northward through Mare
Tranquillitatis. The only other ridge on the moon
to rival it is in Oceanus Procellarum close to
the limit on the right. On Plate 4–5 the beautiful
little crater Bessel is at the terminator. A day and
a half later (Plate 4–7) the whole of the floor
of Serenitatis is illuminated and Linné is seen
as a tiny bright spot not far left of the gap con-

45

necting this mare to Mare Imbrium. Just beyond the gap are the ringed plains Aristillus and Autolycus. At this phase the great mountain ranges bounding the mare are especially conspicuous because of the long shadows they cast in late afternoon.

In the southern part of the moon one of the roughest areas of the whole surface can now be seen. The terminator enters on it near phase five days, and does not complete the passage until almost a week has passed. It is probable that this great area is characterized by some of the oldest of the extant lunar features. When the terminator

(Plate 4–3) is near that gigantic "grandfather Janssen, the shadows of its low walls are lon and render it conspicuous. Each day it become less noticeable. During the many hundreds o millions of years (perhaps billions) since its ge esis, younger features formed over it have begu to obliterate this venerable ancestor. On Plat 4–7 a still grander patriarch is conspicuous o the terminator. This monster is so shallow that was almost completely neglected until rece years, when the geologist Spurr called attentio to its unusual character. Belatedly the Intern tional Astronomical Union voted to give it th

Plate 4–5. Lick Observator 1946 August 4ᵈ 04ʰ 04ᵐ U phase 6.67 days, colongitu 348°. Terminator at Maurol cus, right-hand part of Ma Serenitatis.

e Deslandres. After two or three days it van-
s from sight. Then, except for the feature
wn as "Cassini's Bright Spot," there is noth-
 special to call attention to the area until the
 shadows of sunset bring it into relief again
bout the time of waning half moon.

The monthly changes in orientation of the
r axis show well on Plates 4–11 and 5–2.

With Plate 4–6, at phase 8.32 days, the right
t of the moon first comes under scrutiny.
vius, a rival of Janssen, is barely hidden near
southern end of the terminator. Maginus is
 to its north. Still farther north and exactly
the terminator is a smaller crater, less con-
uous than the larger ones in the southern

region. Nothing about it, no matter how careful
the observation, would indicate that in a very
few days it will excel all other craters and rival
the maria for attention. This is Tycho, possibly
the seat of the most violent lunar explosion of
which there is visible record. Nothing of its great
ray system now appears nearby, but far to the
left, where the sun already is high, its rays have
begun to appear. The study of violent explosive
craters and their ray systems furnishes clues to
the story of lunar evolution.

The great mountain-walled plain Ptolemaeus
and the craters Alphonsus and Arzachel form a
chain that is seen at its best close to the terminator on Plate 4–7. As soon as the sun becomes

Plate 4–7. Lick Observatory, 1940 December 08^d 04^h 19^m UT, phase 8.80 days, colongitude 14°. Terminator at Tycho, right of Deslandres, and Eratosthenes. This plate and the preceding one illustrate an effect of the libration in longitude.

Plate 4–8. Lick Observatory, 1940 August 13^d 04^h 16^m UT, phase 9.33, colongitude 28°. Terminator right of Clavius, near Copernicus and Sinus Iridum.

Plate 4–9. Lick Observatory, 1942 July 23^d 04^h 36^m UT, phase 9.69 days, colongitude 31°. Compare Mare Crisium with Plate 4–8 for the effect of libration in longitude.

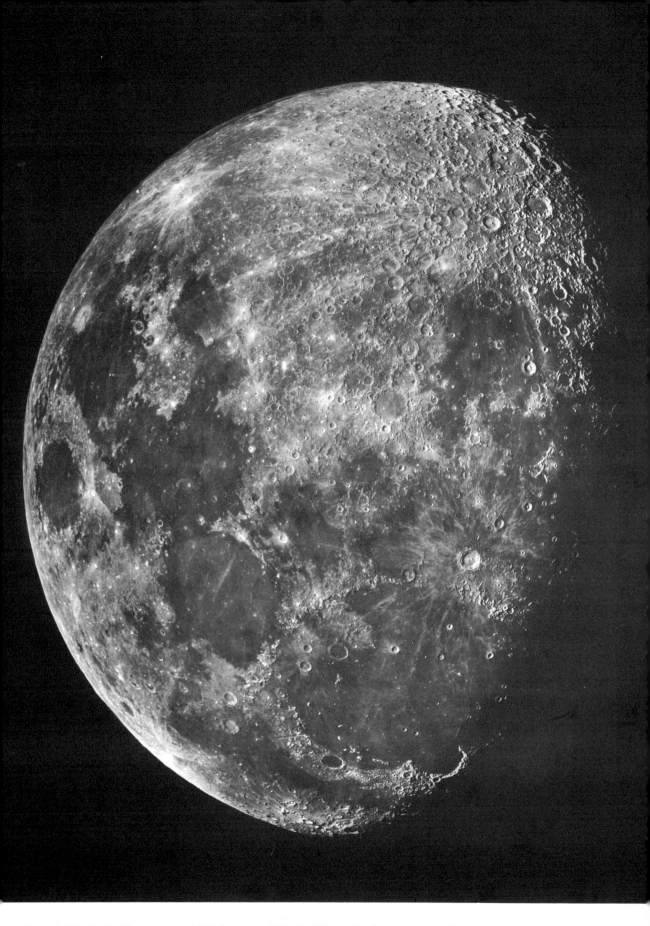

Plate 4–10. Lick Observatory, 1938 January 12^d 03^h 55^m UT, phase 10.37, colongitude 38°. Terminator at Mare Humorum, left of Kepler, and right of Sinus Iridum. Perhaps this is the best plate of the Moore-Chappell series.

high (Plate 4–11), these objects will almost disappear. At full moon phase the easiest way to recognize them will be by use of the famous tiny black spots on the floor of Alphonsus. These spots blacken more and more as the sun gets higher, or at least they assume such an aspect in comparison to the neighboring areas. It is from within Alphonsus that a very small amount of extreme low-density gas was observed to escape a few years ago. Also on this plate the terminator almost exactly on the Straight Wall, an 80-mile long cliff roughly a thousand feet high. That cli

Plate 4–11. Lick Observatory, 1947 May 3^d 06^h 30^m UT, phase 12.09 days, colongitude 62°. Surface of moon beginning to take on the flatter appearance of the full moon phase. Note extreme counterclockwise orientation of the lunar meridian.

me into existence when the surface to the right it collapsed during the formation of Mare Nuum. No other single feature on the moon, with e possible exception of the Altai Scarp, exhibits clearly the manner of mare formation. Farther the north the Apennines, grandest of all lunar ountain ranges and the one that most nearly sembles terrestrial ranges, are resplendent in e early morning sunshine. The scarp of these ountains, the greatest at least on this side of e moon, is seen as a boundary of Mare Imium. On the right hand point of the range the sing sun touches the crest of Eratosthenes. In other day or so it will look like an immense ountain that became too heavy and collapsed ithin itself. (The caldera of Mauna Loa in awaii, which formed because of the reverse ainage of lava, resembles it on a smaller scale.) elow the Apennine scarp we see the pensula and the "marshes" that resulted from llapsed material that was not completely indated when the formation of Mare Imbrium curred.

In the north, close to the terminator, are the vo great ringed plains Archimedes and Plato. hese craters are old, and each was modified by e formation of Mare Imbrium. To the left of lato are the Alps and the Alpine Valley, the ost famous of its kind on the moon. Perhaps hen the terminator is in the position of Plate -7, at colongitude 14°, more interesting features e strung out near it than at any other time. At is phase the left hand portion of the floor of nbrium shows more fine detail than at any her phase until far into the waning stage shown map on Plate 3–2. Sections of two photoraphs have been enlarged as Plates 10–19 and 0–20 to tell a partial story of Imbrium and its rigin. No one yet knows the whole story of this rgest of the maria, although some astronomers e confident that certain parts of it differ from ny of the others.

Plate 4–8 for the first time reveals Copernius, which has the advantage of standing in a ather smooth area and is therefore conspicuous om the moment sunlight first strikes its outer ft hand wall. Its ray system reveals fully as ong a story as does the system of Tycho. The range oval rays to the left of Copernicus show icely on Plate 4–10. The observer with even a oderate-sized telescope can see some of the short, elementary components of the major rays, especially in the mare area to the north. Seen with a more powerful telescope some of them display craterlets at the ends which are pointed toward Copernicus.

At this phase (Plate 4–10) the long, low ridges of Mare Imbrium are observable near the terminator, in the right hand portion of the sea, provided one uses a rather large telescope with fairly high magnifying power and on a night when seeing conditions are quite steady. Some of these ridges are more than a hundred miles long, but they may not be more than fifty feet high, so they disappear as soon as the sun is well above the horizon. At the same time the observer should look for craterlets on the part of the floor where the sun has barely risen. Some of them appear to be rounded, suggesting an origin in the collapse of such domes as may be seen in some of the smooth lunar areas.

The sun rises on Sinus Iridum, the Bay of Rainbows, between the colongitudes of Plates 4–9 and 4–10. This scene will well reward long observation. Study of the craterlets on the highlands just to the north is important in any consideration of origins.

In Plate 4–11, about two days before full moon, Mare Humorum has just emerged into the sunlight on the southern part of the moon. Its shoreline is especially rich in "half-craters." On its northern shore great Gassendi exhibits a lowered seawall that is actually broken at one spot. Much farther north Aristarchus, the brightest of all craters, is on the terminator. The sunlight is just beginning to cross Oceanus Procellarum, which is far larger than Mare Imbrium. It too would be called a mare if its vast size did not so greatly exceed the total area of the maria. The area south of Aristarchus and to the right of Kepler has come into good observational illumination. Much of it, especially in the general neighborhood of Marius, is unusually rich in domes and ridges and such areas should be studied carefully whenever the terminator is near them.

At this point the moon has begun to take on the appearance that is generally characteristic of full moon phase, a period that may be considered as extending from two days before until two days after the actual full moon. The next chapter deals with this four-day interval.

53

5 THE FULL MOON

At full phase the moon is more beautiful to the naked eye than at any other time. One reason may be that at the moment of its rising in the evening it appears far larger than when it is riding high above the horizon. Actually, its angular diameter at this time of its greatest apparent magnitude is smaller than it is at any other position, because the distance to the moon along a tangent to the earth's surface, as at rising or setting, is approximately 4,000 miles more than the distance along the center-to-center line when it is overhead. The reason for this well-known moon illusion is quite interesting and perhaps not widely understood by the non-scientific public.

The human eyes can estimate the size of an object, or its distance or displacement in space, up to a distance of about two hundred feet. Any estimates of size beyond that point are the result of familiarity with the thing which is observed. We know, for example, that the distant house is not merely a doll's house close at hand. Our mind tells us that it is far away although our vision cannot give even a rough estimate of the distance.

It follows from this that when the moon is far from the horizon, and there are no nearby objects with which to compare it, we automati-

cally and unconsciously bring it in toward t[he] 200-foot distance, and it appears as a small o[b]ject. (Indeed, few people realize how small t[he] moon actually is in our sky; a dime held at a di[s]tance of a little more than six feet will exact[ly] cover it.) But when the moon is rising or setti[ng] it is obviously farther away than some dista[nt] house or other large familiar object along t[he] same line of vision, and we judge it to be by f[ar] the larger of the two. This illusion results fro[m] our training and cannot be experienced by [a] small child, who may instinctively reach out h[is] hands to grasp the bright object. This difficul[ty] in measuring distance is compounded by anoth[er] factor. The inability of the eye to make estimat[es] of distances may handicap a person who is n[ot] used to clear mountain air. Accustomed to ha[ze] between himself and comparatively nearby o[b]jects, he tends to allow for the haze in his es[ti]mates of long distances. Langley, a great nin[e]teenth-century astronomer, related in his bo[ok] *The New Astronomy* his experience while hea[d]ing a solar eclipse expedition in the Sierra N[e]vadas. "The sky is cloudless, and the air so cle[ar] that all idea of the real distance and size of thin[gs] is lost. The mountains, which rise in tremendo[us] precipices above him, seem like moss cover[ed] rocks close at hand, on the tops of which, here and there, a white cloth has been dropped; but the 'moss' is great primeval forests, and the white cloth large isolated snow-fields, tantalizing the dweller in the burning desert with their delusive nearness."

Another very interesting phenomenon in connection with the moon's appearance is the effect of refraction, the bending of light by the earth's atmosphere. Refraction causes all objects to appear to be closer to the zenith than they actually are. At the horizon, where the line of vision passes through more of the atmosphere, the effect is rather large—fully a half-degree. For example, when the lower limb of the sun or moon is apparently exactly on the true horizon, the whole of the object is actually entirely below the horizon. As one looks even half a degree above the horizon, refraction becomes noticeably less. As a result of this lessening with altitude, the lower limb of the sun or moon actually at the horizon appears to be lifted toward the upper one, causing the object to appear distinctly oval, with its vertical diameter too small.

Full moon occurs at the instant during the month when the earth is most nearly in line between the sun and moon. In those months when the moon is closely enough in line north and south, an eclipse occurs. Usually the moon passes a little to the north or to the south of the earth's shadow.

At the instant of full moon, a man on the moon and at the center of the moon's visible disk would see the sun directly overhead, and would cast no shadow. Even if he were at the limb, where his shadow would be long, the shadow would be behind him and silhouetted from any view on earth, because the sun would be directly behind the shoulders of an observer on earth. The full moon shows us almost none of its black shadows.

All through a clear night the full moon provides enough light for ordinary vision, yet a light faint enough to soften and romanticize every object. The yellowish tinge of the light, caused primarily by the earth's atmospheric absorption, increases the effect of mystery and subtlety. In selecting a time to visit an observatory a non-astronomer usually requests a night of full moon, but the sight is usually disappointing because the great shining disk is almost devoid of the shadows that define lunar features.

To the naked eye, the full moon exhibits smoothish, dark areas, the maria, and brighter, mountainous regions. The pattern of the maria gives us the traditional "man in the moon," which is apparent at full moon because of variations in the small percentage of light reflected by different terrain. The walled plains and craters that were so striking during the waxing phase, and will be again in the waning moon, owe their visibility to shadow patterns. When the sun is low the long shadows of their walls exhibit spectacular detail, but of course at full moon they almost disappear. In contrast to these objects that hide from us at full moon, there are other craters that are the result of almost unbelievably violent explosions that modified the nature of the lunar surface rock. Such craters, and also the maria, become much more conspicuous under a high sun, when the extreme roughness of the surrounding territory has been minimized by lack of shadows. For example, Tycho brightens far more than any of its neighbors as the sun gains altitude. Long and rather bright streaks called "rays" extend from it in various directions for hundreds of miles (Plate 5–1). Only the greatest of the maria can rival its display at this time.

To a conspicuous but lesser degree the same thing happens to Copernicus, Kepler, Byrgius, Olbers, Aristarchus, Anaxagoras, and others. Even craterlets may show such rays conspicuously as, for example, the two just outside of Furnerius and Stevinus. Some craterlets brighten until at full phase they appear almost as stars against the moon.

The pattern that crisscrosses the full moon is a glorious one but the glare dazzles the eye of the observer at the telescope too much for him to perceive any real detail. With a low-power eyepiece to permit simultaneous viewing of the entire full moon, one can become lost in a mysterious, alien, shining, and dreamlike world of almost hypnotic appeal. But the serious investigator cannot devote much of his time to such reveries, and his studies force him to work usually at other phases. If, however, he has superior equipment, if he has lived with the moon for a long time, and if devising methods of observing the moon has become a kind of game for him, he eventually learns that there are some important things revealed by the full moon that do not appear at the other phases. Even use of a cardboard diaphragm to reduce the brightness of the image by stopping down the objective or the mirror of his telescope, or the insertion of a filter at the eyepiece, materially increases what may be observed. Partial elimination of glare is a first step toward visual study.

But elementary procedures of this kind do not improve the situation sufficiently to permit much valuable research today on the full moon. There is no way by which the astronomer can do much to overcome the lack of sufficient visual contrast. Fortunately there are other, non-visual methods that can provide assistance. The chief of these is the photographic plate. The observer can choose from a wide variety of plates and may expose and develop them in ways to obtain less contrast or to exaggerate contrast as much as he likes. He can choose plates and filters to produce photographs that appear at their brightest in any desired color. He can record the scene as it is transmitted by radiation of entirely different wavelengths from those that produce recognizable images on the human retina. He can choose plates so sensitive that they can preserve for his leisurely perusal a record made in a mere thousandth of a second. Whereas a visual observation is subject to the prejudices of the astronomer and leaves no record to display to a skeptic, what is clearly recorded on the negative is fact; only interpretation can be challenged. The human eye directly at the telescope has one advantage: When the photographic plate shows some tiny markings and cannot quite resolve them, the astronomer can use a visual telescope on a night when the air is steady to look at the area and observe a little more of the delicate detail than was recorded by the camera.

The application of photography is so vital to a comprehensive knowledge of the lunar surface, and especially of what it displays at full moon, that a digression outlining the functions and techniques of lunar photography may be useful at this point.

The first celestial photograph was a daguerreotype of the moon made in 1840 by Dr. J. W. Draper. It was of rather poor quality. Ten years later Dr. W. C. Bond, first director of the Harvard Observatory, made an excellent daguerreotype with the fifteen-inch refracting telescope of that observatory. Soon after these were taken, and as a result of their exhibition in Europe, Warren De la Rue of England became so much interested in celestial photography that he devoted the remainder of his life to it. In 1853 he made some exquisite lunar photographs, using the newly developed "wet plate" process and a thirteen-inch reflecting telescope.

By 1879 the "dry plate" process had been invented, with much advantage to every sort of photographic enterprise. During the 1880's the use of fluorescing dyes in plate emulsions extended the spectral range toward the red. The earlier plates had been very little affected by light of longer wavelengths than those of the green, and their color balance was therefore quite different from that of the human eye, which is most sensitive to the yellow wavelengths. By 1906, with the advent of "panchromatic" emulsions, red was finally able to contribute its full share to an exposure. Nineteen years later the use of still other dyes produced plates sensitive to "infrared" light, that is, to wavelengths too long to affect the human retina. Since that time the improvement of infrared plates has been steady and today we have a rather wide range of such plates. Their speed has been much improved, varying degrees of contrast are available, still longer wavelengths are recorded; fineness of grain, which affects resolution, has been increased.

De la Rue had used a reflecting telescope because of an unavoidable defect inherent in the refracting type. Refraction is bending. What a prism does is to bend the shorter wavelengths of light, such as violet, more than it does longer ones like the red, and thus spreads white light into a spectrum. But when a lens exhibits a similar effect there is trouble. The first telescope (three centuries ago) used only one piece of convex glass as an objective. Such a telescope brings violet light to a decidedly shorter focus than it does the red, thus distorting the image— when one color is brought to a sharp focus all other colors contributing to the image are out of focus. More than two hundred years ago Dollond lessened this trouble considerably by combining two pieces of glass of different density and different surface shapes so that each would tend to correct the defects of the other. Telescopes with such objectives are called *achromatic refractors*, although the correction is not perfect. But the loss of light involved in use of a multiple lens system in a large telescope would be very serious. The reflecting telescope, invented by Isaac Newton, brings all colors of light to ex-

ctly the same focus and is therefore free from
nis difficulty.

It is, nevertheless, possible to make excellent
ictures of the moon with a refractor if a yellow
lter is placed in front of the photographic plate.
Lenses of telescopes made for visual observa-
ons are so shaped that color difficulties are
ss in the yellow range than they are in either
ne blue or the red range. The filter excludes all
ne light except the range which it was designed
» pass.) All but one of the pictures used to
lustrate Chapter 4 were made in this manner
ith the 36-inch refractor of the Lick Observa-
ry and a special yellow filter. Of course, the
se of a filter increases the necessary length of
ne exposure. The reflector, on the other hand,
oes not require a filter, except for special effects,
id is therefore faster than a refractor of the
me light-gathering power and focal length. It
possible to "take" a picture of the moon in
ss than a thousandth of a second, although most
xposures are of the order of a few tenths of a
cond. Many of the pictures used in this book
ere made using the 60-inch reflector at Mt.
Vilson with infrared plates. In this study a filter
hich passes no visual light was used with them.
This way the unsteadiness of the earth's atmos-
here distorts the images formed by infrared
ght less than those formed by blue and violet.

When we prepare a full moon picture for
ublic display, the dense parts of the negative,
hich picture the bright areas, may be chemi-
lly reduced until they compare more nearly
ith the parts that show the dark maria. Other-
ise the bright areas may show merely a glare
white paper or the dark maria may be shown
; an almost useless black. This has been done
xpertly for Plate 5–1, and the result is one of
ne finest full moon pictures possible for public
cturing and for illustration of popular articles.
owever, the part of a plate that has been over-
xposed and then reduced can never possess the
ll quality of a plate that has received proper
xposure and development. Furthermore such
mpering renders the plate subject to suspicion.

Many of the details of Plate 5–1 are lost in
egatives of the full moon not treated in this
anner, as well as in viewing the full moon di-
ctly with a telescope. Peppered with craters and
cking an atmosphere, however, the full moon
ill shows many more details than can be picked
it in Plate 5–2 of the full earth (north at the

top), photographed from an Apollo spacecraft.
Here the vast differences between the earth and
its satellite stand out strikingly.

The earth's oceans appear black in Plate 5–2,
with none of the detail of the lunar maria. Clouds
obscure the view of earth in many regions, and
the earth's cloud cover averages 50 percent. Much
of Africa can be picked out in Plate 5–2, however, as well as the large island off its southeast-
ern coast, the Malagasy Republic. The Asian
mainland is on the horizon toward the northeast.

Without an atmosphere and water, and consequently weather, the moon offers a very different
landscape from the earth. All meteoroids crash into the moon, but all except the largest are burnt up
in earth's atmosphere. Although many meteorites
have blasted large craters on the earth over the
eons, water, wind, and weather have quickly
erased most of them. But stark craters billions of
years old still pock the lunar surface. Ghost craters there were nearly overwhelmed with flooding
lava and debris from larger meteoroids.

In Plate 5–1 of the moon, made with a high
sun but with proper exposure and choice of plate,
Copernicus shows surprising facets. Plate 10–22,
showing Mare Imbrium, is almost ugly because
of the white expanses of the bright areas, but
it is one of the most useful plates ever made for
the study of the floor of that mare. The student
who looks at a few fine pictures is apt to feel
that he "knows" the appearance of the moon. In
reality he is like one of the nine blind men who
examined the elephant: He knows a few aspects,
and there are many. The observer who desires
really to know the moon must continually experiment with new and tedious techniques. Often
he learns after several nights of hard work that
some approach for which he had high hopes is
valueless. On the other hand, other new techniques—perhaps the use of polarized light—will
have merit. The radiation-sensitive cells of the
photometer could be usefully applied for a thousand hours or so to provide charts of the relative
brightness of a hundred features under all sorts
of conditions. Kozyrev has recently surprised
lunar scientists by demonstrating that the spectrograph is needed. When each of these techniques has been pushed to the limit, the group
of "blind men" who have used them may compare notes and as a result may learn something
of the true appearance of their "elephant," the
moon.

Plate 5–1. Lick Observatory, 1946 January 17^d 07^h 51^m UT, phase 13.80, colongitude 82°. This is a beautiful plate taken about two-thirds of a day before full moon. The plate has been specially printed to avoid the extreme full moon contrast. At full moon many of the details are lost in the glare.

Plate 5–2. Apollo 17 astronaut's photograph of the full earth, made in December 1972, on return trip at close of Apollo Missions of the National Aeronautics and Space Administration. View extends from Mediterranean Sea and Arabia (top) to nearly all of Africa (center) and Antarctica (bottom).

6 THE WANING MOON

The preceding chapters on the waxing and the full moon described the aspects of the moon with which amateurs are most familiar, and for an obvious reason: those are the phases that are observable in the evening sky. The waning moon is another matter, and a much less convenient one. Even one day after full moon the moon does not rise until almost an hour after sunset. One week later it rises at midnight, and at the end of the phase it withholds its appearance until dawn. The amateur who observes systematically under these conditions is of no ordinary breed.

Just before full moon, objects in the sunrise area very close to the right hand limb reveal their detailed structure best. Just after full moon much the same is true for the features close to the left hand limb, which we now see almost as plainly in the sunset as when the waxing moon was a thin crescent. Only the visual impact of the brightness of the rest of the moon interferes. On the photographic plate the images are equally good. However, the reversed direction of the shadows from the setting sun does help interpretation very much when we compare the two views.

At phase 16.3 (Plate 6–1) the chain of ver large craters northward from Furnerius throug Petavius, Vandelinus, and Langrenus can be see plainly. It is almost sunset on them. Actually w can usually see this chain better after full moo than we could on the waxing crescent becau the latter is observed close to our horizon, whe the earth's atmosphere nearly always interfere On the waxing crescent the inner left hand wa of the craters were dark and the right hand on were brilliantly illuminated; now the conditi is reversed. The observer who wishes to look f rills, craterlets, and other fine features has deta to observe different from those he had two wee before.

Now Mare Crisium revises the story it to during the early phase. The floor still shows t same lacy detail, the shores have the same ba and the shape is unchanged (unless the libratio have changed the foreshortening by a shift the distance of the mare from the limb). B now the second brightest feature of the mo makes its appearance just off the right ha shore, attracting almost as much attention as t mare itself. This is the extremely bright explosi crater Proclus. Craters of this kind, which ha esulted from an unusually violent explosion, brighten under a high sun much more than do he surrounding areas. (Even Tycho is not conspicuous when the sun is near the horizon.) When the sun was rising on Mare Crisium it was even lower on Proclus. Now, when the sunset terminator just touches the left hand shore of he mare, the sun still is rather far from the horizon of Proclus. Once more the observer should look at the peculiar ray system of Proclus and should compare the right hand gap of the system with the similar one in the greater system of Tycho. The rays cannot be seen to the right where apparently a partial sinking produced a "swamp" called Palus Somnii, the marsh of sleep. Its low, rough surface is seen most clearly on late 6–2.) There is a sharp boundary on the ght between the swamp and Mare Tranquillitatis.

At approximately this phase of 16 days Plate 6–1) the great ray to the left from Tycho to Mare Nectaris can be studied best. It should be noted that it is actually two rays and not one, being doubled over much of its length, and that most of its brightness comes from craters along

.

Plato, in the northern area, is always interesting and at this phase the craters and partial craters around its rim can be observed to good advantage under a high sun. The contrast between the rim and the dark floor is beautiful. The amateur with a fairly large telescope is apt to find himself lured, as so many have been before him, into a specialization that will continue throughout the remainder of his life. Does or does not Plato sometimes exhibit a haze over its floor, visible when the sun is low? Perhaps this ringed plain has been observed more than has any other single feature on the surface of the moon.

Toward the limb to the right, Aristarchus now shows its peculiar relationship to the large disturbed area adjacent to it on the lower right. Beyond that area Seleucus can be seen with its peculiar long ray which, in its branch to the left, becomes the northern boundary of that disturbance. Just right of Aristarchus is the tortuous Schroeter's Valley, the widest of all rills, and nearby is the crater Herodotus.

Plate 6–1 shows Olbers clearly, almost at the limb. If Olbers were not at the limb it certainly would be fully as conspicuous as is Kepler. Because of its location, however, Olbers can be observed only when the librations are unusually favorable, although it can be located at other times because of its ray system. The long ray from Seleucus and a ray from Kepler extend to it (Plate 6–2). Any person who wants to know the nature of lunar rays must become familiar with the oversystem of Copernicus-Kepler-Olbers-Aristarchus, and Seleucus. Possibly the double upper right hand ray from Tycho extended to Kepler and tied in with this oversystem at one time but later had its northern end obliterated by Oceanus Procellarum.

To the south of Olbers and a little farther from the limb is Grimaldi, a small mare with a floor that is one of the darkest areas on the entire satellite. South of Grimaldi is Byrgius, another very bright explosion crater, which would be much better known if it were near the center of the disk.

On Plate 6–2 the disk has lost most of its full moon appearance. Once more Eratosthenes is conspicuous, although Copernicus has not yet faded from its noonday grandeur. The number of observable starlike points scattered over the lunar surface is now smaller because of the lowering altitude of the sun. Ptolemaeus once more is starting to change its appearance toward the wonderful object it was ten days ago.

Between Plates 6–2 and 6–3 Mare Nectaris is at its evening best. On both of these plates the Altai Scarp is dark and the less rough surface between it and Nectaris shows nicely. These plates would give one of the best possible representations of the ridges on the floor to the right if they had been printed to a lesser density, but this of course would have spoiled the detail of the brighter parts. (Plate 10–8 shows such lighter shade in an enlargement.) In the north on Plate 6–3 the terminator is on Posidonius and the time is right for a study of the ridges and craterlets on the floor of Mare Serenitatis that complements the observations made at the time Plate 4–5 was exposed. Turning far to the right we find this to be as satisfactory a time as any to observe the famous right-angle triangle made by the rays of Copernicus, Kepler, and Aristarchus. The plumelike, elementary components show quite well. The serious amateur should observe ray structure at every opportunity and watch the

Plate 6–1. Lick Observatory, 1938 December 8^d 06^h 56^m UT, phase 16.29 days, colongitude 104°. The sunset terminator has almost reached the left shore of Mare Crisium. Farther south Furnerius and Stevinus are beginning to emerge from the glare of light which made observation of their forms very difficult.

Plate 6–2. Lick Observatory, 1937 October 22^d 08^h 03^m UT, phase 17.84 days, colongitude 122°. The left part of the surface has lost the full moon appearance. The right part of the disk still resembles the full moon phase.

Plate 6–3. Lick Observatory, 1937 October 24^d 13^h 01^m UT, phase 20.05 days, colongitude 149°. The Altai Scarp to the left of the sunset terminator shows dark. In the north the sunset has reached Mare Serenitatis.

Plate 6–4. Jet Propulsion Laboratory. The waning earth from about last quarter (left) to waning crescent (middle) and nearly new earth (right), as photographed from the moon by Surveyer 7 spacecraft. North is at the bottom here as for all the photographs of the moon in this chapter.

anges in it that take place as the height of the
n changes.

In all of his observing the watcher should
bitually make notes whenever he is certain
any peculiar detail. If he has the proper skill
should make drawings of all special features,
mparing them with what he can see under
fferent lightings. Much that he has observed
ll prove to be illusory. In my own experience
e suggestive power of a visual illusion was
monstrated by observations of the central peak
Eratosthenes. After a series of afternoon ob-
rvations I was once certain that there was a
aterlet at the top of the peak; my craterlet
rsisted for several days. I soon found, how-
er, that as the direction of the light changes
om right to left the craterlet completely dis-
pears. It is all a trick of the shadows. One
od visual observing practice is to trace out
e major features in pencil on tracing paper
aced over an enlarged photograph. The paper
en is placed on white cardboard and details
e filled in while the observer is at the tele-
ope directly observing the features.

Plate 6–4 shows a series of photographs as
en from the moon by Surveyor 7 spacecraft
ade of the waning earth to compare with the
aning moon. The waning of the earth (from
ft to right) as seen from the moon is the reverse
the moon's phases as seen from the earth.

On the right hand floor of Serenitatis, at
phase just beyond that of Plate 6–3, Linné is
ain at its best. The pit craterlet observable
waxing half moon now can be seen again at
e top of the ten-mile white mound. Its appear-
ce is entirely different from that under the op-
site lighting.

At these phases when Linné is best observ-
le, the Caucasian and Apennine sections of the
brian scarp sparkle in the late afternoon sun-
ine. Their appearance at this time should be
ntrasted with what it was in the early morn-
g. Also in the Alps, to the north of Mare Imium, the Alpine Valley shows nicely. At this
ase the row of craterlets along the valley's
rthern rim is a fine feature to observe with a
ir-sized telescope. The craterlets along the
uthern rim appear most advantageously under
moderately low morning sun. The difference
phase for best observation of these rows prob-

ably is due to opposite slopes of the ground outside the valley.

Phase 22 days (Plate 3–2) demands that the
visual observer examine Clavius and Maginus for
their rim craters, Tycho for detail of its floor,
Deslandres, Alphonsus, Ptolemaeus, the craters
of the Apennines, the radial valleys outside Aristillus, the left hand floor of Mare Imbrium and
the ringed plain Plato. All of these features are
lined up on the sunlit side of the terminator.

By 24 days (Plate 6–5) the waning moon
has become merely a broad crescent that rises
around 2 in the morning. At the terminator in
the north the Bay of Rainbows (Sinus Iridum)
has begun to disappear into the night. Aristarchus
has lost enough of its dazzle to permit observation of the general form of the crater. If conditions are just right the small mountain at the
center of its floor may be seen, and there may
be a hint of craterlets where the floor joins the
inner wall.

On the upper right hand portion of the
moon Mare Humorum is at its afternoon best.
Just to the left of it and closely paralleling the
shoreline, the three "Clefts of Hippalus" are
rather easy to observe. On the left hand portion
of the mare floor several ridges are easily seen.
A half dozen craters, varying from conspicuous
to almost unobservable, demonstrate their relationship to the ridges. Several scattered floor
craters look very black at the centers of whitish
spots. A continental shelf can be seen on the
right floor. On the northern shore Gassendi is
a splendid object. Its southern floor has sunk, but
not sufficiently to have caused degeneration to
a mere bay. These features can be more easily
observed on the enlargement, Plate 10–5.

South of Mare Humorum the great walled
plain Schickard may be observed well when the
librations are favorable. Some observers have
thought it should be considered as a small mare,
but the general shade of its floor does not match
that of the maria. To the left of Schickard we
can see Schiller, the most eccentrically shaped
of all the craterlike formations. The moon's limb
in this general area should also be observed because, whenever the librations permit, some high
mountains may be seen projected along the limb
at this point.

The craters on the extreme right hand portion

Plate 6–5. Lick Observatory, 1938 August 20^d 12^h 47^m UT, phase 24.37 days, colongitude 206°. The sunset terminator is crossing Sinus Iridum and has almost reached Kepler and Mare Humorum. Grimaldi shows dark near the right limb.

Plate 6–6. Lick Observatory, 1936 Septemb[e]r 12^h 57^m UT, phase 26.40 days, colongitude The sun is setting on Kepler. Aristarchus, close terminator, has lost sufficient brightness to easy observation of details.

of the floor of Mare Imbrium and those much farther south, where the floors of Mare Nubium and Oceanus Procellarum join, are as important as anything else that can be examined at this or at a later phase. When the moon rises shortly

before dawn, the terminator is moving to th[e] right across these floors, illuminating craterle[ts] and ridges until the crescent shown on Pla[te] 6–6 is reached, about three days before ne[w] moon.

COPERNICUS THROUGH THE MONTH

e features of the moon testify in general to a
atively quiet period of formation as compared
the geologic activity that has apparently taken
ace on the earth. This generalization applies
most of the mountain-walled plains such as
eslandres, Ptolemaeus, Clavius, Bailly, Riccioli,
d Grimaldi, all of which tend toward a single
t or shade over most of the surface and are
erefore very conspicuous only when the sun
low and casts long shadows; it applies to some
e craters that are observable under a low or
moderate sun but that practically vanish at
ontime.

There is, however, a good-sized minority of
lent craters which, although they occupy only
small percentage of the lunar surface, dominate
e appearance of the full moon. Such craters
em to indicate explosions far greater than
ything of the kind for which evidence remains
the surface of the earth. Among the most
mous of this type are Copernicus, Tycho, Kep-
, Aristarchus, Olbers, Proclus, and Manilius.
naller craters of this kind are rather common,
d near full moon bright craterlets are num-
red in the hundreds. Each of them indicates
e site of a very violent local explosion.

The diameter of the floor of Copernicus is
about 40 miles and that of its narrow rim is 56
miles. The mean height of the rim is perhaps
12,000 feet, although the rim height above the
external plain is much less than it is above the
inner surface. (It must be remembered, however,
that the degree of uncertainty as to heights and
depths on the lunar surface is far greater than
one would surmise from a study of the photographs.) Along the rims there are peaks that
depart widely from the average level.

Copernicus is conspicuous even when the
sun is low, as shown by Plates 7–1 and 7–2. Its
shallowness in comparison to its diameter is
shown by the fact that the floor, which is entirely dark in Plate 7–1, is almost completely
illuminated in Plate 7–2 despite the fact that
there is not much difference in altitude of the
sun. If we wished to compare Copernicus to a
pie pan nine inches in diameter, its depth would
be only a third of an inch. When the sun is
high, Copernicus exhibits entirely different characteristics from those which it reveals near
sunrise and sunset. Plate 7–4, made at nearly full
moon, reveals a very different appearance of Copernicus and the lunar landscape around it than

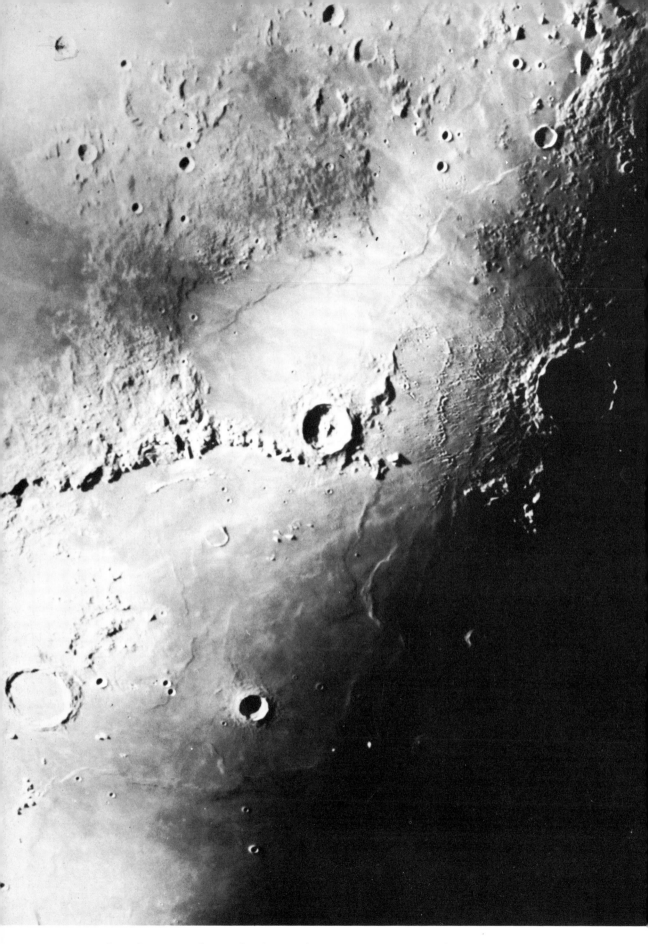

Plate 7–1. Mt. Wilson and Palomar Observatories, 1956 November 12^d 04^h 04^m UT, phase 9.47 days, colongitude 20°. Sunrise on Copernicus.

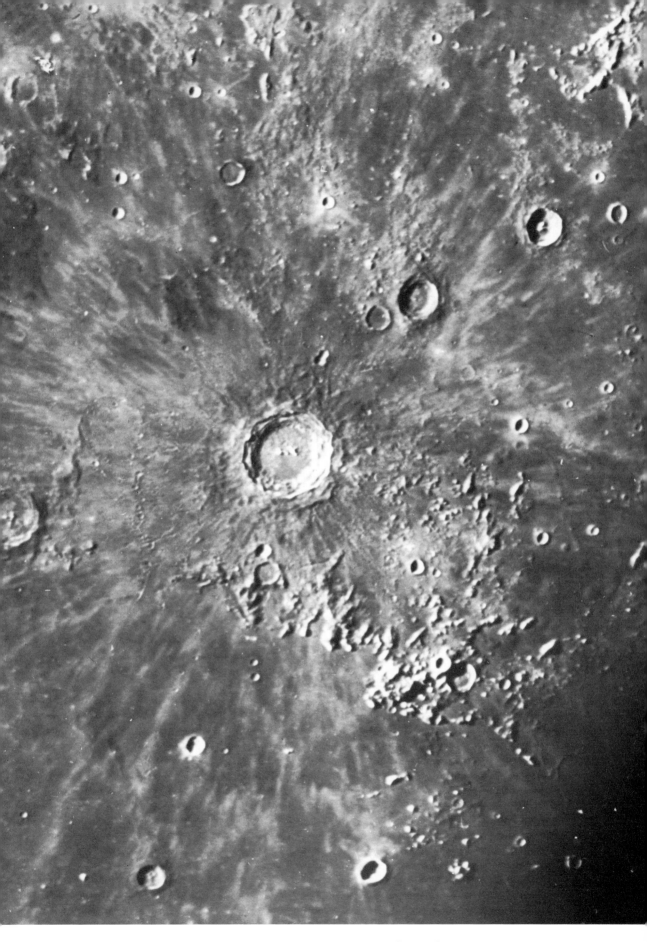

Plate 7–2. Lick Observatory, 1938 July 7^d 04^h 23^m UT, phase 9.30, colongitude 24°. With Plate 7–1 demonstrates shallowness of Copernicus.

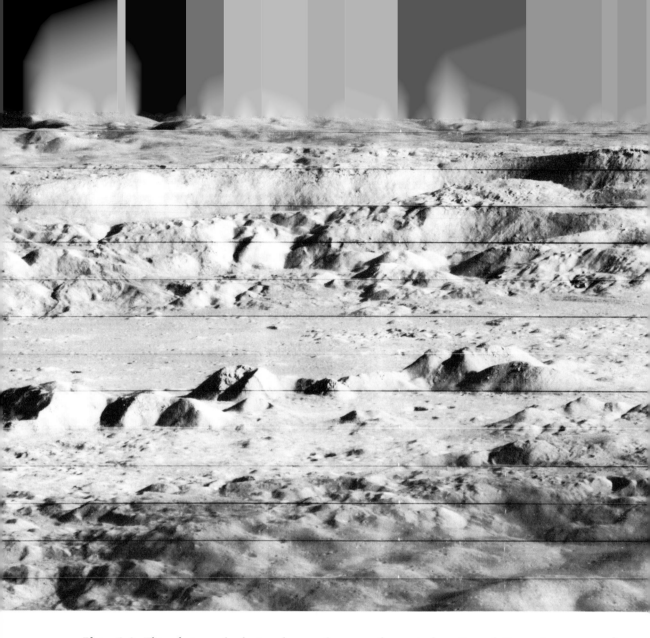

Plate 7–3. This photograph, facing due north across the central portion of Copernicus was made at 0:05 U.T., November 24, 1966, using the telephoto lens of Orbiter II. The Orbiter cameras and their techniques are described briefly in Chapter 19. Because of the expanse of sky, this picture, unlike the others in this book, is printed with north up. Orbiter was about 48 km above the surface and roughly 250 km south of the center of Copernicus. The east-west diameter of the part shown is approximately 30 km and is a bit less than a third of the crater's diameter. The rounded mountain in the upper left-hand corner is the ringed plain Gay Lussac. Comparison with Plate 7–2 shows, on each, the row of isolated central mountains. Each picture also reveals that the floor to the north of that row is smoother than it is to the south. The projection of the southern rim of Copernicus against the floor causes it to stand out in sharp contrast. The northern rim, very roughly 100 km farther away, is much less conspicuous. A long, wide, east-west valley can be observed plainly in the northern wall, crossing most of the picture. In even the best-made photographs from the earth's observatories, the valley appears as a mere terrace near the upper part of the wall. Probably it is a graben, formed during the subsidence which modified the original form of the crater.

Lunar students, almost unanimously, believe that at least the initial form of Copernicus was due to the impact of a small asteroid, or giant meteoroid. This excited internal forces which brought about a series of caldera-like collapses that enlarged the floor and left the wall terraces as scars. (See page 117. But although Apollo astronauts have explored many kinds of lunar terrain, they have not landed near such large craters as Copernicus, and it will take years to work out their full history.

es Plate 7–2, in which the sun was low. Plate
2 clearly shows the great ray system around
pernicus, as well as the details of its slumping
lls and inner floor, and highlights of its cen-
l mountains. These last can be seen much more
early in the closeup photograph of Copernicus
ade by Lunar Orbiter II in Plate 7–3. In this,
e could be standing near the south rim of the
ater, viewing its floor, its central mountains,
d its northern wall as one prepares to explore
Plate 7–4 was underexposed in printing so that
e brightest parts of the surface would not be
urned out," thus revealing the actual appear-
ce of Copernicus when the sun is high.

Plates 7–2 and 7–4, different as they are,
ch give incomplete but equally valid repre-
ntations of the crater. Shadows are responsible
r the general appearance of the first, and the
portance of low features is exaggerated in it.
his exaggeration is often not recognized as one

studies a beautiful photograph, but such recognition and the proper use of it are essential to all
serious study. Observations made under a high
sun, on the other hand, scarcely hint at such
differences of height but, for such craters as
Copernicus, they do reveal beautifully the peculiar brightening that develops as the altitude of
the sun increases. These high-sun photographs
are not as crisp and exciting as those with long
shadows, but they are equally important in
interpreting the lunar surface. The same principles hold true for the corresponding visual observations.

Plate 7–5 is an intermediate print, made
when the sun was moderately high, and exposed
and printed to show best the fine structure of the
ray system of Copernicus. The many short elementary rays, with their plumes extending radially away from Copernicus, and the complex
structure of the great rays, are all vital data for

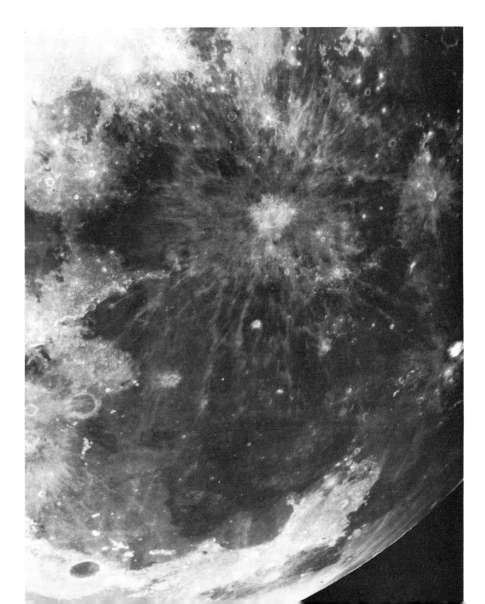

ate 7–4. Mt. Wilson and
lomar Observatories, 1958
ly 1ᵈ 09ʰ 27ᵐ UT, phase
.06 days, colongitude 87°.
ort exposure to show detail
Copernicus under a high
n.

an understanding of the complex phenomenon that is Copernicus.

These changes in appearance of Copernicus through the month are almost unbelievable to the novice. It is mainly the existence of such cyclic changes, which are characteristic of many other features also, that has caused the moon to be the favorite of the amateurs.

In Plate 7–1 (colongitude 20°) the whole of the floor of Copernicus and even the tops of the central mountains are dark. The sunlight just strikes the upper part of the inner right hand wall. The outer wall to the left shows terraces, radial valleys, and craterlets.

On the flattish surface to the left of the wall, this sunrise phase begins to show the famous patterns made by rows of confluent craterlets, made conspicuous by long shadows despite their small rise above the general level. At rare and fleeting moments, under the best possible observing conditions, observers using powerful telescopes have reported seeing thousands of such tiny craterlets, especially on the left hand slope.

Among those who have witnessed this phenomenon have been George Ellery Hale, W. W. Campbell, and H. P. Wilkins. Those of us who have observed the moon consistently for a long enough time to have experienced such fraction of a second can never forget these startling flashes of vision. More than anything else they reveal to us how little we know of the fine detail of the lunar surface, despite several centuries observation.

On Plate 7–2 the altitude of the sun is still low but already most of the floor of Copernicus is illuminated. The three central mountains show as well as at any other time of the month. The shadows exhibit conspicuously the difference between the northern and the southern halves of the floor. Even in the low sunlight, the northern part is seen to be quite smooth, while the southern half is a mass of low hills. The structure of the inner wall becomes apparent at this phase and will unfold more and more as the altitude of the sun gains moderately. The three famous terraces of the inner wall can be seen. Some

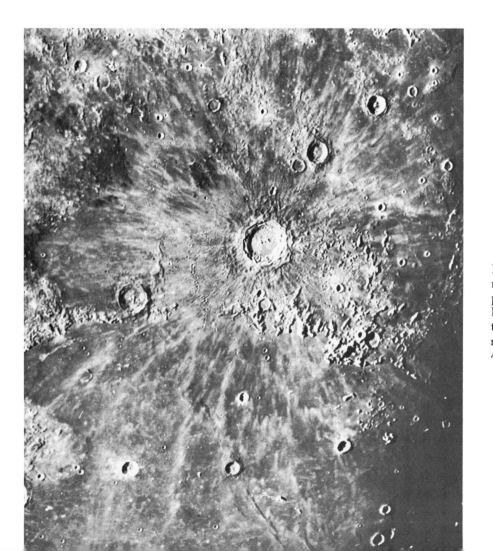

Plate 7–5. Detail of Copernican ray system. The ring plain, Eratosthenes, is the large circular depression the left and just below Copernicus, in the center. See Plate 4–10 for technical details.

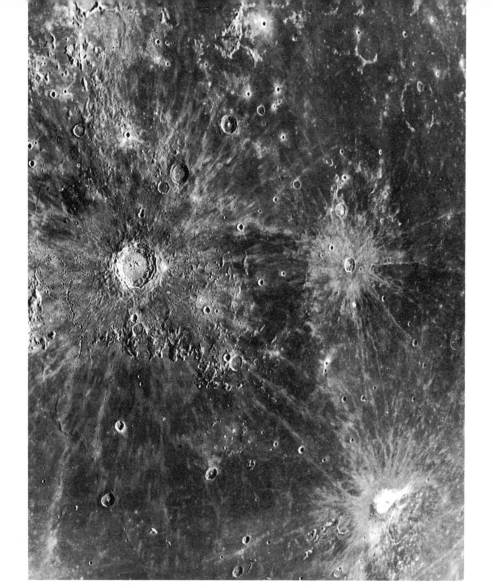

ate 7–6. The Copernicus-
epler-Aristarchus oversys-
m of rays. See Plate 3–2 for
chnical details.

aters appear on the wall. As the lighting
anges, different details will in turn reveal their
orphological characteristics. The difference in
eepness of inner and outer walls is evident at
glance. The inner wall to the left is still dark,
it the outer wall to the right is already bright.

On Plate 7–1 everything right of Copernicus
as still hidden in night-time blackness. On
ate 7–2 everything for fully a hundred miles
the right is bathed in low sunlight. The result-
g long shadows show lines of craterlets, mostly
dial to Copernicus but not with the pronounced
tterns found to the left. Hortensius is observ-
le to the right of Copernicus and to its north
a group of very low, rounded domes.

These domes are not observable except when

the sun is rather low. Many of them, perhaps all,
show a tiny, black crater pit at the top. Farther
to the north, beyond Milichius, is a dome larger
than these of the Hortensius group which shows
the crater pit very distinctly. A smaller dome is
found quite close to the crater. The frequency
and distribution of domes is one of the newer
studies of selenography.

It can be assumed that the rays of Copernicus begin to appear half a day after sunrise, although it is almost impossible to set a time with
complete accuracy because the density and contrast of even good photographs affect visibility
very much. The rays increase their relative
brightness until sometime near noon and then
hold it for much of the afternoon, fading out be-

fore sunset. The best time to observe their fine detail is under a moderate sun. Plate 7–6 was made at the time of the waning half moon. It was printed in an attempt to show the details of the ray system to best advantage. A simultaneous study of it with Plate 7–5 will show many peculiarities in the ray system of Copernicus. One is the fact that the major rays are not radial to Copernicus. The second is that in Mare Imbrium, north of the crater, there are many plume-shaped short rays which are radial to Copernicus. The points of the feathers are toward that crater. In a few cases a craterlet is observable on the pointed end of such an elementary ray and in nearly all cases a brightish spot can be seen there that can be assumed with some confidence to contain a craterlet. Examination of the two major rays extending north-

ward into Mare Imbrium shows that they ha[v] a complex structure. Despite overlapping, the[re] are places where this structure can be observe[d] as composed of the elemental plume rays, whi[ch] are radial although the complex rays are n[o]

To the left of Copernicus a peculiar featu[re] known as Stadius can be observed, a circle [of] craterlets elevated very slightly above the ge[n]eral level. Adjoining Stadius on the right is [a] less regular but similarly enclosed area, and ru[n]ning northward from it is the most conspicuo[us] line of small craters on the moon. The li[ne] continues directly into the strongest of the ma[jor] rays of Copernicus. On the south of Stadi[us] (Plate 7–5) and touching it are the two mo[st] peculiar of all rays. They are ovals, one insi[de] the other. Examination of these ovals shows th[at] they, like Stadius, are loci of craterlets. T[he] plumed elemental rays are found througho[ut] their area and other rays, generally radial, e[x]tend outward from the perimeters of the ova[ls]

On the upper right, the Copernican ray sy[s]tem is much complicated by the additional pre[s]ence of rays from Kepler. Featherlike comp[o]nents belonging to each of the craters are obse[rv]able.

Plate 7–7 was made just before sunset a[nd] reveals the same sort of detail as was shown [in] the early morning. Eratosthenes, which had a[l]most disappeared under the high sun, is alm[ost] exactly on the terminator and is as conspicuo[us] as it was at sunrise. The hexagonal shape of t[he] rim of Copernicus shows nicely at this pha[se] contrasting with the nearly circular shape of t[he] floor. The contrast between this picture of t[he] rim and that of Plate 7–4 is one of the mo[st] startling to be observed on the moon.

The Copernican phenomena which ha[ve] been described are the principal ones to c[on]sider in any study of the nature and origin [of] Copernicus. A satisfactory hypothesis may n[ot] necessarily demand the existence of all of the[m] but it must not forbid any of them. The da[ta] for Copernicus and for all other similar crat[ers] must be more or less formally combined bef[ore] any valuable study is possible. Even then t[he] hypotheses of candid and able selenologists a[re] contradictory. These disagreements may pers[ist] even after lunar explorers have actually visit[ed] the site; after all there is still disagreement ab[out] many of the features of the earth to which ge[ol]ogists have relatively easy access!

Plate 7–7. Mt. Wilson and Palomar Observatories, 1955 October 9^d 11^h 58^m UT, phase 23.24 days, colongitude 188. Late afternoon on Copernicus.

TYCHO THROUGH THE MONTH

frequently happens that a person looking at the full moon through a telescope for the first time exclaims "Look at the north pole!" What prompts the exclamation is the view of Tycho and its rays as shown by Plates 5–2 and 8–3. It looks amazingly like the polar circle with radiating lines of longitude. Even an observer who knows perfectly well that the poles of planets are not marked by any external signs and that meridians are artificial lines drawn for convenience on maps is struck by the similarity.

Tycho and Copernicus are competitors for pre-eminence among the craterlike features of the lunar surface. When the sun is low Copernicus definitely outclasses all other ringed plains, but when the sun is high Tycho is so magnificent that even the great maria scarcely rival it. Copernicus has the advantage of being situated in one of the dark, smooth maria. Tycho is in one of the roughest bright areas of the moon. Around it are great mountain-walled plains (Plates 8–1 and 8–2), some of them with diameters more than twice that of Tycho. When the sun is low the shadows cause these walled plains to stand out in majesty. The surfaces of nearly all the walled plains are composed of rock of one shade, and when the sun is high and shadows are short they almost disappear. But at such times Tycho

and its brother ringed plains brighten far more than the surrounding areas, and divert our primary attention from the other features.

In Plate 8–1 Tycho is exactly on the sunrise terminator. Only the crater rim is illuminated; all of the interior part, including the central mountain, is in darkness. Crater rims always are higher than are any central mountains that may exist. The rough features to the left of Tycho are emphasized by their long shadows. To the north of Tycho, toward the bottom of Plate 8–1, is Deslandres, the largest of all the mountain-walled plains, and one that is certainly older than most of the other craterlike enclosures on the moon. On this plate it is observed as one of the most conspicuous of all features, but its walls are so low that at even a day's phase later one must look carefully to find it. Plate 8–2 shows within Deslandres an intensely bright spot that is related to the Tychonian ray system. This spot is bright, however, even when it is near the terminator with the sun too low for Tycho's rays to be observed. It is called "Cassini's Bright Spot" because of the assertion by that astronomer, three centuries ago, that he had observed a cloud there and that a short while later a "new formation" first became visible. It is interesting that both Neison and Goodacre insisted mistakenly that

Plate 8–1. Sunrise on Tycho and Deslandres. See Plate 4–6 for technical details.

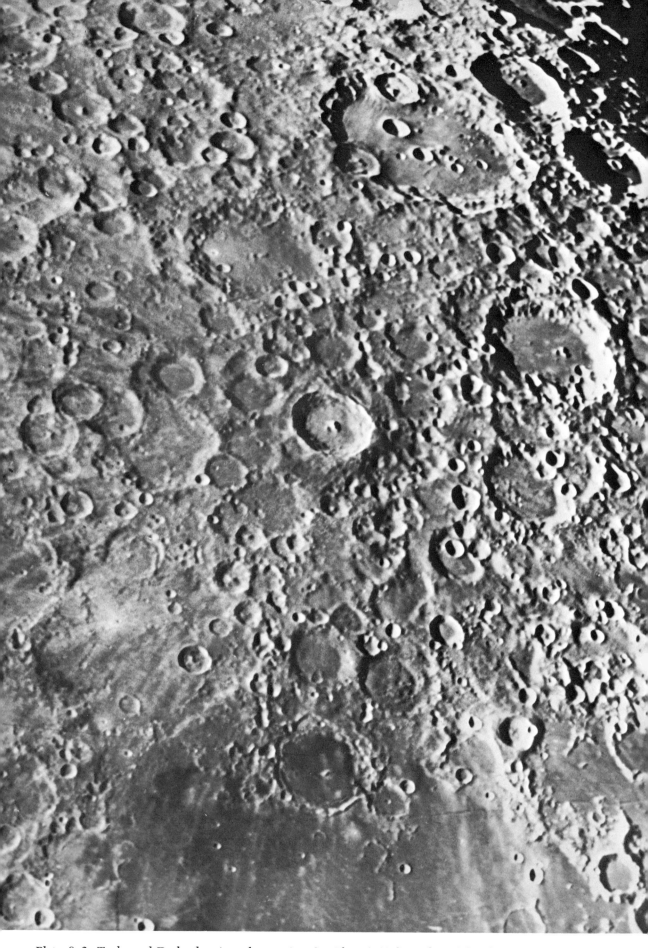

Plate 8–2. Tycho and Deslandres in early morning. See Plate 4–10 for technical details.

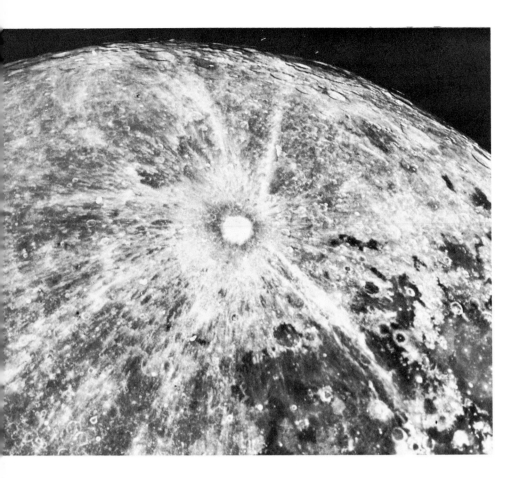

Plate 8–3. Details of r[..]
system near Tycho. S[..]
Plate 5–2 for technical d[..]
tails.

the feature does not exist. A small crater (Plate 8–2) is easily observable throughout a good part of the lunar day. Undoubtedly Cassini was mistaken concerning the seventeenth-century origin of the object, but the later students must have made insufficient observations, and those at the wrong time of the month. This is a very easy error for even an experienced observer to make concerning features about which he has only casual interest. The bright spot is useful in locating Deslandres when the sun is high. Plate 8–8, made in the afternoon, exhibits the craterlet almost as beautifully as does Plate 8–2, made in the morning.

Plate 8–2 was made at a phase only a day later than that of the preceding picture, but a significant change has taken place. All of the floor of Tycho is illuminated, giving proof that, as in the case of Copernicus, its depth is small in comparison to its diameter. Among the mountain-walled plains Deslandres has become almost invisible and changes in others, such [as] Maginus, are noticeable. The point to rememb[er] is that the mountainous areas of the moon a[re] in general less rugged than observations near t[he] terminator would indicate. Tycho is much l[ess] deep in relation to its diameter than even [a] shallow pie pan, and the walled plains such [as] Maginus are still flatter. A man standing at t[he] center of Clavius or of Ptolemaeus would ha[ve] no idea that he was even within an enclosure b[e]cause of his point of view.

On Plate 8–3, made five days after sunr[ise] on Tycho, the whole area is entirely unreco[g]nizable. (Plate 10–4, which has been correct[ed] for foreshortening, well may be studied sim[ul]taneously with Plate 8–3.) It is possible at t[his] phase to make exposures of negatives and pho[to]graphic prints in such a way that to some [ex]tent they restore to visibility the details of [the] central portion of Tycho, but mere exposu[re] cannot restore the lost mountain-walled plai[ns]

hich depend almost exclusively on shadows for
eir visibility. At first glance Plate 8–3 shows
ycho as a circular white blob with great, al-
ost radial, bright spokes. It may be observed
xt that the image of Tycho is scalloped, where
aters and parts of craters exist on its rim.
nmediately outside the rim is a dark area into
hich the rays scarcely intrude. Neison wrote
 this area as a plateau but an examination of
rious photographs indicates that it is more
ely a slight depression. Nine of the ray spokes
ist if we consider the double one toward the
wer right as two and also include the rather
definite one that bisects the great gap to the
ght. Between the major rays, we find the same
rt of elemental plume rays as around Coperni-
is. Unlike the major rays, they are radial. The
ajor ray toward the lower left corner of Plate 8–3
especially notable. Even a casual examination

will show that it ends before the edge of the
picture and is replaced by a closely parallel ray
that starts about half-way between Tycho and
the edge. Plate 8–4 has been made, by use of
a high-contrast emulsion and proper choice of
exposure time, to show the structure of this ray,
which is primarily a series of bright craters. A
fuller discussion of this ray belongs properly to
a study of the nature of the lunar rays.

Plates 8–5 and 8–6 are enlargements of a photograph of Tycho and its surroundings made after
the intensity of the great rays has decreased in the
late afternoon sunshine. A partially sunken oval
area breaks the wall slightly at the lower right-
hand portion of the rim and therefore cannot be
older than Tycho. Plate 8–7, made fairly early in
the morning by Lunar Orbiter V, reveals much
more of the detail of the walls, floor, and flanks of
Tycho. Some of the numerous domes on the floor

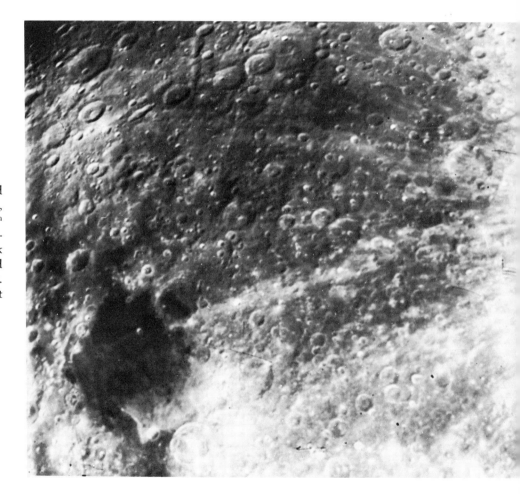

ate 8–4. Mt. Wilson and
lomar Observatories,
)58 August 1ᵈ 07ʰ 22ᵐ
T, phase 15.53 days, co-
ngitude 105°. Kodak
ᵀ–N plate with infrared
ter for extreme contrast.
etails of Tycho's greatest
ft-hand ray.

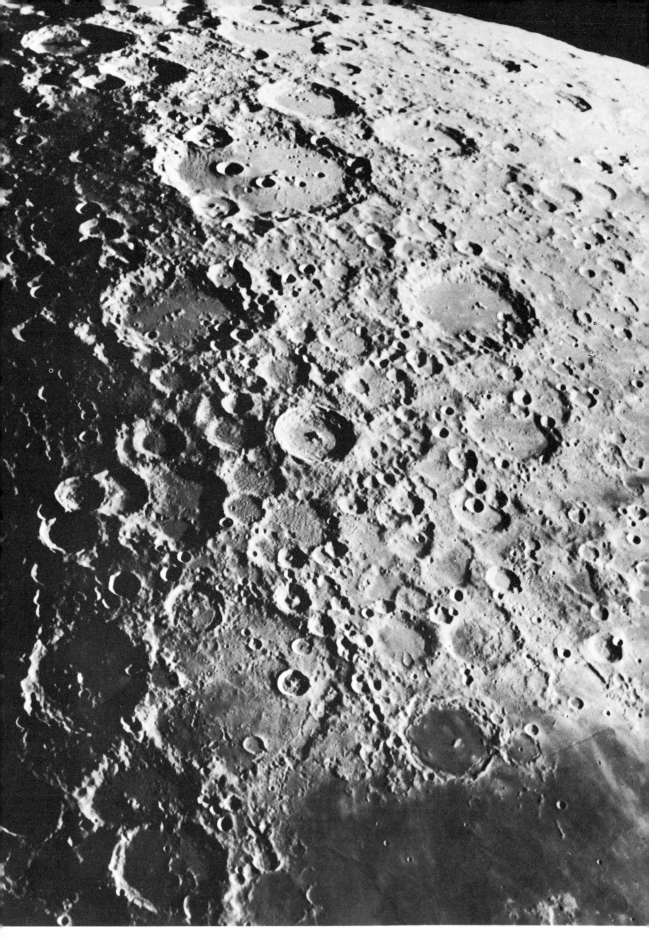

Plate 8–5. Mt. Wilson and Palomar Observatories, 1955 October 8^d 12^h 53^m UT, phase 22.27 days, colongitude 176°. Southern area of moon at third quarter.

Plate 8–6. Enlargement of Plate 8–5 to show Tycho in late afternoon.
See Plate 8–5 for technical details.

Plate 8–7. Tycho, photographed from Lunar Orbiter V from an altitude of 220 km, is an example of young, fresh craters. Tycho's rim is hummocky, its walls are slumped, and the floor rough, with many domes and fissures. Tycho is about 85 km in diameter. Scale: Framlet width = 7 km.

ate 8–8. Fading of Tycho's
ys. See Plate 6–2 for tech-
cal details.

ppear to have summit pits and may be volcanic
ones or fissures. Molten lava also seems to have
elled among the slumping walls at the south
f the crater. Plate 8–8, made under a moderately
igh afternoon sun, has been printed to empha-
ize the narrow bright rim of Tycho. At both
he inner and outer edges of the rim the slopes
ppear to begin so suddenly that they indicate
sharp breakaway. At various places both within
nd outside the bright ring can be seen the rim
raters that are an almost universal accompani-
nent of the large craters, whether or not they
re sites of violent explosions. The detail of this
icture is different in almost every way from
hat of Plate 8–5. It would almost appear that
eatures of entirely different shape had replaced
nost of those shown on the latter plate. As in

the case of Copernicus the reason is rather obvious: Plate 8–5 depends mainly on shadows and
roughness but Plate 8–8 depends primarily on
variations in the reflectivity of the surface. The
two plates correlate quite well insofar as the
truly explosive features are concerned; the quies-
cent features, on the other hand, display almost
no correlation.

Plate 8–9 advances the time to sunset, with
the central peak casting its long shadow all the
way to the inner left hand wall. Tycho is no
more conspicuous than it was at sunrise, but
during the day its radical changes in appearance
have furnished data that are valuable for interpreting not only the origin of the violently explosive craters but also the lunar surface as a
whole.

Plate 8–9. Mt. Wilson and Palomar Observatories, 1958 July 9^d 10^h 40^m UT,
phase 22.11 days, colongitude 185°. Setting sun on Tycho.

SUNSET ON PTOLEMAEUS

he frequent observer of the moon, whether rofessional or amateur, is generally enthralled y the fast-changing panorama of shadows at inrise and sunset. The lack of atmosphere on ie moon is the cause of the striking contrasts t these times. There is almost no portion of the inar disk where light and darkness mingle and egate each other; each point is either in full inlight or in blackness relieved only by reflec- ons from neighboring elevations. The sky is tally dark and there is no twilight.

The shadows of mountains, ridges, and low levations unobservable except under angular ght carry much valuable information. The four hotographs reproduced here as Plate 9–1 illus- ate the rapidity of the change in appearance. ine would expect that the changes would be ow; after all, the sun shines on any point on the ioon for more than two weeks at a stretch. But ie interval between the first and last of these ictures was only 79 minutes, a short enough me for an amateur to observe continuously. he largest of the craterlike formations is the exagonal walled plain Ptolemaeus. Its floor is pproximately 90 miles in diameter and ordi- arily it is considered very flat, except for the rater Lyot, which is conspicuous. Alphonsus is ie smaller feature above Ptolemaeus. Walled

plains are depressed areas in mountainous re- gions. Typically they exhibit little or no explo- sive action. They are not truly surrounded by walls but rather by the rough mountains of the area. There are often numerous high peaks on these mountains but at some places there are great gaps.

In the first picture with the setting sun still illuminating most of the floor the shape of the shadows indicates a gap in the right hand rim of Ptolemaeus. Even in the last photograph, where the sun scarcely touches the floor, there is no shadow at all at the gap; the break in the wall is obviously complete. Above the gap (south) is a very complex mountain mass with the ap- pearance of an almost unbelievably sharp, high peak. However, much of this steepness is merely the illusion caused by the exaggeration made by shadows from a low sun. During the short in- terval of time represented by this sequence of photographs the tip of the shadow of this peak moves to the left across the floor, until in the last picture it has merged with the blackness that has engulfed the left hand portion of the plain. This general darkening to the left is caused by the curvature of the moon's surface, which shows nicely on the central plain.

On the floor of Ptolemaeus nearly two hundred tiny craters are observable under proper conditions. Almost none of them shows in these pictures. In addition to these craters there are perhaps three dozen fairly large but extremely shallow depressions that are usually referred to as "saucers." Only three or four of them can be seen when the sun is high, but in the grazing sun they stand out in considerable detail, as they do in these pictures. In the upper right hand portion of the floor is an almost perfect circle of nine of the saucers. Seven of the nine can be observed in the first picture, but by the time the last picture was taken nearly all of them were hidden by the shadow of the wall.

Examination of the area lit by sunlight streaming through the gap in the rim makes it clear that progressively more and more of the roughness of the floor becomes visible as the su nears the horizon. Not only are the saucers see but there are ridges and valleys, generally tan gent to the saucers. Such features have too littl elevation to cast shadows except when the su is practically on the horizon.

Students of the moon are uncertain of th cause of these shallow depressions. Perhaps the are "ghosts," remnants of old-time craters tha came into existence soon after Ptolemaeus too form, while its floor was still so plastic that the were almost destroyed by the subsidence of the material. More probably most of them are ghos of old craters which existed here before Tych came into being. Whatever the cause may hav been, it was apparently an important factor i forming some of these great walled plains.

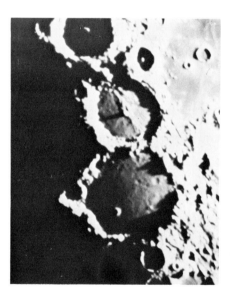

Plate 9–1. Mt. Wilson and Pa omar Observatories, 1955 N vember 7^d 11^h 38^m UT; 12^h 04 12^h 57^m; phases 22.67 to 22.7 days, colongitudes 181° to 182 Sunset on Ptolemaeus.

Plate 9–2. Ptolemaeus (below and Alphonsus (above) as the appear with the sun at an ang] of 70°, photographed by Lun Orbiter IV from an altitude 2719 km. Many craters pepp the floor of Ptolemaeus. Th famous dark spots and clefts Alphonsus are clearly visible. fine crater chain appears to th right of Ptolemaeus and anoth appears on its right rim.

ie origin of the maria, like so much else in con-
ction with the mysterious satellite, is contro-
rsial. The hypotheses fall roughly into three
asses, with minor variations and crossovers.
ich of them is supported by able students and
iyone who wishes to know the moon must be
miliar with all three.

The three types of hypotheses are the fol-
wing:

1. The maria were caused by collisions with
teroids, the molten rock which produced the
ther flat surfaces resulting from the collisions.

2. The maria may have resulted from such
illisions but the molten rock, at least in part,
tisted previously on the moon.

3. The molten rock came from beneath the
irface of the moon as a result of great surface-
ide convulsions. If collisions with asteroids did
cur they did no more than trigger those con-
ilsions. If the convulsions were of a tidal na-
ire they may themselves have produced at least
irt of the molten rock.

Any hypothesis concerning the formation of
ie maria must explain—or at least not contract—the following known facts about the seas:

1. The surfaces of the maria approximate

flat plains more closely than do any other large
areas of the moon.

2. The maria are the largest of all lunar features; some are visible even to the naked eye.

3. The maria are much darker than the general area of the lunar surface. Only smooth floors
within certain craters and a very few small mountainous areas reflect so little light.

4. The larger maria are interconnected.

5. Long, very low ridges are common on the
surfaces of the maria. Two of the systems are
more than 800 miles long and are significantly
close to meridional in direction.

6. Some large "craters" and a good many
small ones appear on the surfaces.

7. The boundaries between the maria and
the adjoining rough areas are usually cliffs. Many
of these are very high and some are several hundred miles long.

8. In general, the "seaward" walls of large
craters that touch the shores of maria are either
lower than the landward walls or are entirely
missing.

9. On the floors of most of the maria there are
"ghosts," which appear to be craters that are entirely or almost entirely submerged.

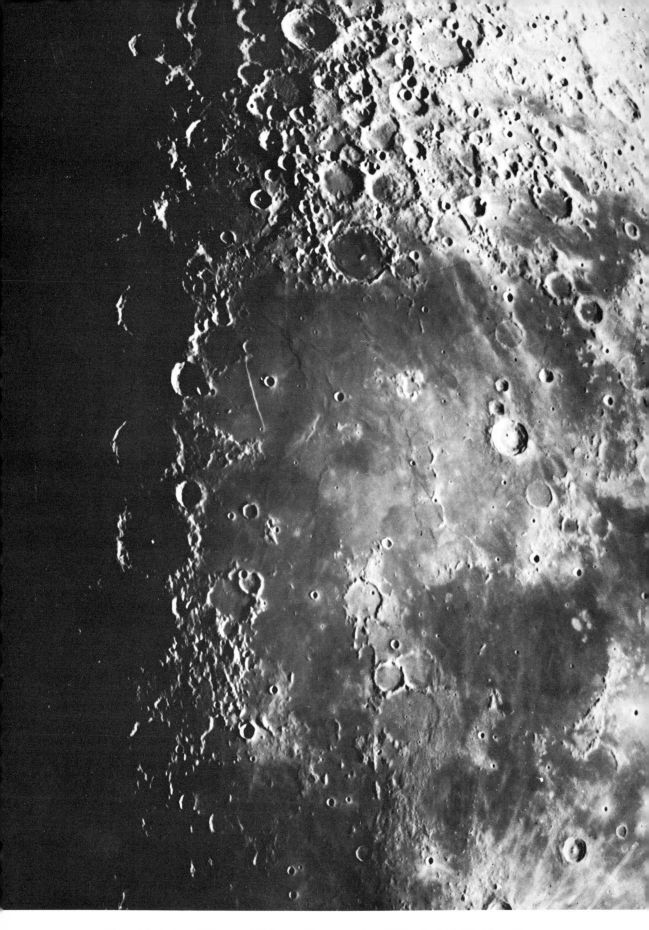

Plate 10–1. Mt. Wilson and Palomar Observatories, 1956 July 30^d 11^h 29^m UT,
phase 22.22 days, colongitude 183°. Floor of Mare Nubium.

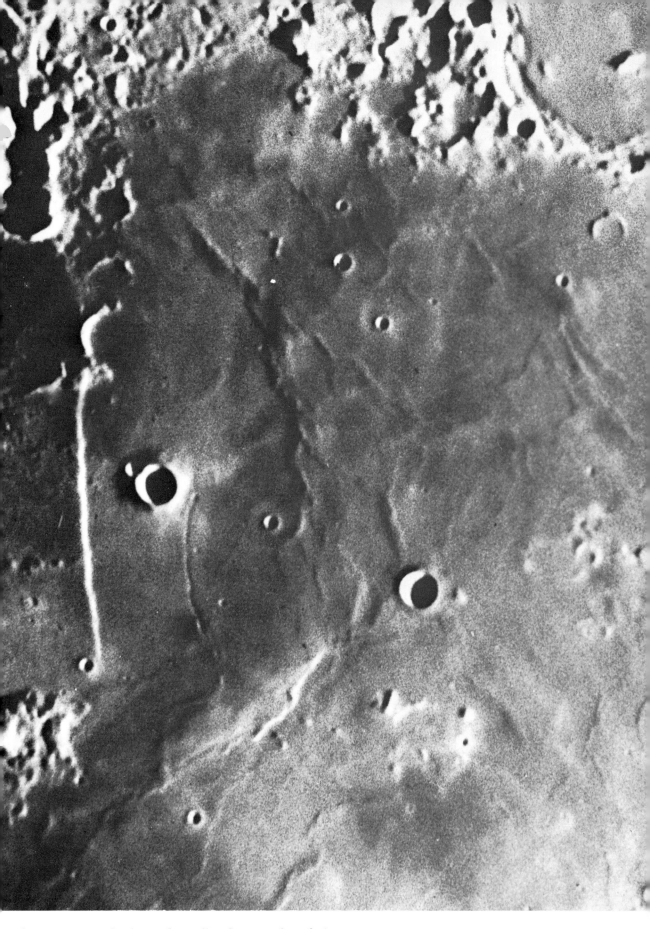

Plate 10–2. Details of Straight Wall and surroundings before sunset.
See Plate 10–1 for technical details.

10. Adjoining some of the maria are large "marshy" areas. These are usually found where the shoreline cliffs are low or entirely missing.

11. "Continental shelves" are common inside the periphery of maria, the floors deepening suddenly some distance from the shoreline. The deeper areas in the floors have sharp boundaries.

12. The northern shores of four of the great maria—Imbrium, Serenitatis, Tranquillitatis, and Foecunditatis—lie very nearly on a great circle of the lunar surface.

13. Four fifths of the maria surfaces on the visible side of the moon are found in the right hand hemisphere.

14. Photographs taken by the Soviet satellite, Lunik II, show that the far side of the moon has little mare surface.

The characteristics of some of the better known and more noteworthy maria are described in the following paragraphs.

Mare Nubium (Plates 10–1—10–3). This mare is interesting chiefly because of the "Straight Wall" and the numerous ghosts fou on its surface. The Straight Wall is a cliff abc 80 miles long and perhaps 1,000 feet high for ing one section of the left hand shore of Nubiu In the afternoon (Plates 10–1 and 10–2) it sho as a bright line but under the morning sun (Pl 10–3) it appears as a dark line. To the left of t wall are seen half of the boundary, and the c responding floor portion, of an old mounta walled plain of huge size. To the right is t mare floor. Under a low sun the almost destroy remnants of the right hand wall are visible. the rough ground to the south of the mare Deslandres (Plates 10–1 and 10–3), another walled plain of nearly the same size as the p tially destroyed one at the Straight Wall. D landres can be observed as an integral featu only when the sun is very low; otherwise its wa do not cast enough shadow to exhibit its for Plate 10–1 also shows the ghost of a third pl of roughly the same size touching the Straig Wall plain on its north. This plain is so nea

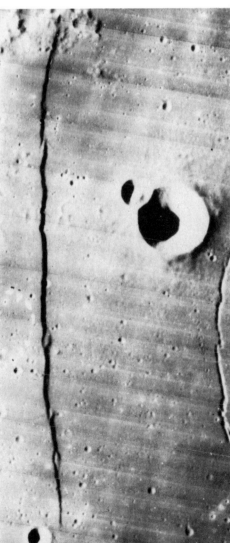

Plate 10–3–A (Far Le Mt. Wilson and Palor Observatories, 1954 vember 4^d 02^h 47^m U phase 8.37 days, color tude 11°. Sunrise on Straight Wall casts a l shadow on Mare Nubiu

Plate 10–3–B (Near Le Made later in the day th Plate 10–3–A, more det of the Straight Wall visible in this closeup p tograph by Lunar Orbi than in the enlargement Plate 10–2. The trough the right involving m small craters is Birt R Birt being the larger of two craters in the cente

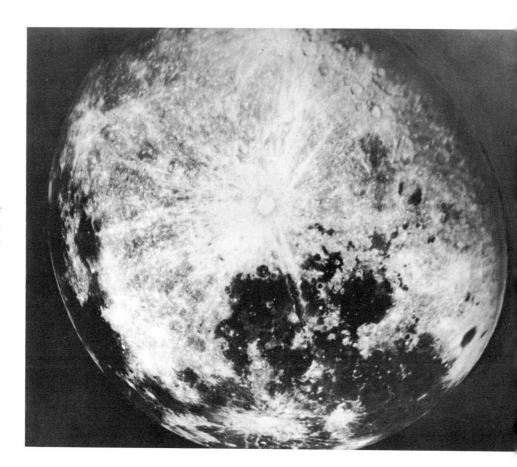

late 10–4. Correction of
oreshortening near Ty-
ho. The technique used is
xplained in the Appendix.

estroyed that it can be recognized only by the
pproximately circular black area. These huge
lains, dim reminders out of the evolutionary
ast, certainly belong to an earlier stage of lunar
istory. At no other place does one find stronger
vidence suggesting that the maria were formed
y a caldera type of collapse of surface area.

On the floor of Nubium (Plate 10–1) is a
ring of three almost engulfed ghosts of old
raters. They extend northward from $x = 87$;
$= 93$ of the map, Plate 3–2. Just to the left of
ie northernmost of the three is a larger ghost,
ie wall of which is almost at the level of the
are floor. A peninsula of high ground runs
outhward from the Copernican area to end in
ie vicinity of these ghosts. Two of the craters
n the peninsula, Fra Mauro and Bonpland, have
lmost lost their walls but probably they have
ot vanished sufficiently to be classed as ghosts.
he crater Pitatus (Plate 10–1) touches the
outhern floor of the mare with the usual result:
s seaward wall is much lower than the rest.

Plate 10–4 is a view of Mare Nubium and
Mare Humorum, corrected for foreshortening by
proper projection of a photograph on a globe.
The method and the story of its development are
told in the appendix. This print is reproduced by
courtesy of Dr. Harold C. Urey. It shows the
two maria and the southern part of Oceanus Procellarum with little distortion. It also shows the
mare-like patches to the south of Humorum.
These areas are closely related to the great gap
on the right in the major ray system of Tycho.

If Mare Nubium in its present condition were
the result of a great explosion caused by either
internal or impact forces it is unlikely that the
ghosts could have remained. There would also
be no reason to expect the low northern wall of
Pitatus and the sinking at the Straight Wall.

If Nubium resulted from a general upwell-
ing, great enough for molten rock to overflow an
area which had undergone a caldera-like sinking,
all of the observed data would appear to be reasonable. This observation of cause does not pre-

91

Plate 10–5. Mare Humorum in late afternoon.
See Plate 6–5 for technical details.

Plate 10–6. Sunrise on Mare Nectaris.
See Plate 4–3 for technical details.

clude the hypothesis that a small asteroid or a large meteorite may have struck near the center of the region and acted merely as a trigger to produce a sinking in an area that was already unstable.

Mare Humorum (Plates 10–4, 10–5). This small sea adjoins Nubium on the right. It is one of the more circular of the maria. The floor shows a distinct continental shelf. Left of the central part of the floor there is a beautiful, irregular, and complex north-south thrust ridge, so low that it is visible only when the sun is near the horizon. It exhibits a relationship to craterlets.

Humorum is notable for its large craters with walls broken at the shoreline. On the left hand shore Hippalus has become a mere bay with a system of three strong parallel rills running from lower left to upper left of it. A continuation of

these rills can be seen beyond a mountainou area directly to the south. They reach Ramsde which itself has a fine rill system that can be see faintly on Plate 10–5.

On the southern shore, as shown by th plate, Lee and Doppelmayer have very badl broken seaward walls. Directly left of Doppe mayer is a fine ghost. Lee is located within th ghost of an even larger crater, so badly broke that it has not been named. On the norther shore of Humorum is Gassendi, one of the mo noted of the ringed plains. Its southern wall lower than the others and at one point is con pletely broken. This mare tells in a less spectacu lar manner the same story as Nubium.

Mare Nectaris (Plates 10–6—10–8). Plat 10–6 exhibits Nectaris and its surroundings whe the sun is rising on the floor to the left. A grea cliff runs southward from Gutenberg to become part of the shoreline on the left. Just inside the floor is one of the most conspicuous continental shelves to be found on the moon. On the southern shore the sun just strikes the crest of Fracastorius, one of the largest of the walled plains. At this phase the very low seaward wall does not show and it appears to be a great bay. Plates 10–7 and 10–8 show Beaumont, another such bay-like crater on the right hand shore. The floor also exhibits a combined ridge and continental shelf inside the shore on the right. At the lower right is Theophilus, a great ringed plain. Plate 10–7 shows that the left rim of Theophilus is lower than the one away from the mare.

The chief interest in Nectaris in connection with the evolution of the lunar surface lies in the Altai Scarp which is far to the right of the shoreline. In Plate 10–7 the rising sun strikes the crest of the scarp, which extends for several hundred miles like an arc centered on Mare Nectaris. This plate shows a partial parallel ridge between the scarp and the mare. There were three major sinkings in the development of Mare Nectaris. Under a low afternoon sun (Plate 10–8) the scarp is dark, proving that it is not a ridge. Both of these photographs show the modification of the surface between the scarp and the sea. They also show the long ridge on the right floor, which is higher on its left than on its right slope.

One might contend that some scarps, such as those of the Apennines, were caused primarily by melting at the line where the break occurred, but such a hypothesis could not possibly apply to the Altai Scarp. Mare Nectaris tells exactly the same story as do Mare Nubium and Mare Humorum.

Mare Crisium (Plates 10–9—10–10). This, the most completely enclosed of the larger maria, resembles the largest of the typical mountain-walled plains more closely than do any of the other maria. A continental shelf parallels almost

Plate 10–7. Sunrise on the Altai Scarp. See Plate 4–4 for technical details.

Plate 10–8. Lick Observatory, 1940 August 22^d 12^h 48^m UT, phase 18.69 days, colongitude 142°. Altai Scarp in the late afternoon.

Plate 10–9. Ghosts and continental shelves on Mare Crisium. See Plate 4–3 for technical details.

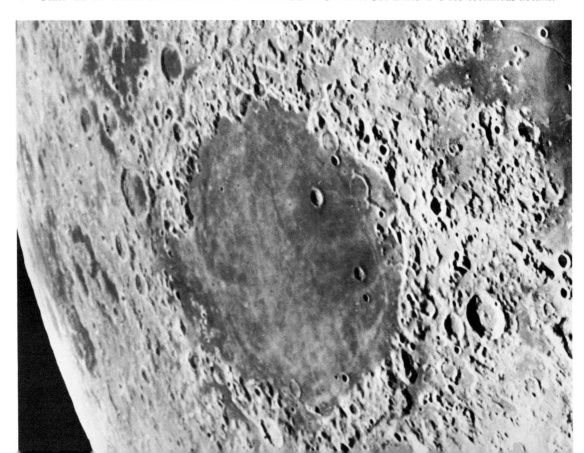

Plate 10–10. Northward oblique view by Lunar Orbiter II of the Marius Hills on Oceanus Procellarum (see Plate 3–24). The hills are named after the 41-km-diameter crater Marius near the upper right corner and believed to be volcanic domes, plugs, and cones. The long wrinkle ridges are part of a 1900-km system of such structures in Oceanus Procellarum. About 230-km stretch of the lunar horizon is shown, clearly curved.

e whole of the shoreline on the left. Plate 10–9
ows, just right of the center, an elongated
nken area that occupies perhaps a fifth of the
or. The "half-craters" on the shoreline of the
are are too numerous and too conspicuous to
quire individual comments. The plate shows
so a lacy network of ghosts, especially on the
rthern part of the floor. Some of them even
st a hint of shadow when the sun is very close
the horizon. At no place do we find evidence
at the plain was produced by an explosion,
ther internal or impact. The floor appears to
ve sunk far below its mountainous surround-
gs and to have been flooded.

The best terrestrially made photographs
ow many domes in the region to the north of
arius. In those pictures they appear as smooth
d rounded, except that many show a craterlet
the top (pages 73, 113, and Chapter 13). Plate
2 shows them nicely in the Copernican area.
the far more open scale of Plate 10–10, all are
en to be rough, often with tiny craterlets scat-
ed over the surface. At the center of this plate
a typical pair. Three have been partially

merged directly below them. In the lower right-
hand corner is another beautiful pair. The upper
left quadrant of the picture is "full" of them. In
the lower right-hand corner, a section of the long
diagonal ridge is seen to be composed of 18 contiguous craterlets. Probably the crater Marius is
a caldera although it may have had an impact
origin. However, a large majority of the other
observable features are due to endoforces.

Since Apollo astronauts did not land on or
near one of the lunar domes, their direct exploration will have to wait for later manned missions.
Perhaps enough will be learned from the analysis
of the materials and data already returned from
the moon to explain how domes form.

The Maria on the Left (Plates 10–11 and
10–12). Both photographs show Maria Crisium,
Marginis, Smythii, Spumans, Undarum, Foecunditatis, and Tranquillitatis, and Plate 10–11 also
shows Serenitatis and Nectaris. Smythii resembles Crisium but is badly distorted by foreshortening even in Plate 10–11. The map of the lunar
earthside in Chapter 19, based on photos made
by Lunar Orbiters, shows the shape and gives

Plate 10–11. The left hand Maria partly corrected for foreshortening (see Appendix).

Plate 10–12. The left hand Maria. See Plate 4–8 for technical details.

us a better understanding of both Maria Smyth and Marginis. The first five of the maria definite appear to be interrelated. Spumans and Undaru are small mare areas which seem to have co gealed before they could completely merge. Th same phenomenon is seen at the upper left ha limb in the case of Mare Australe (Plate 4–11 When inspected casually, the whole area giv the impression of an interrupted subsiden which, if it had been completed, would have pr duced a single mare as large as Mare Imbriu One tends to get the feeling that a gigantic for was acting at once on the whole of the lunar su face, a force which apparently had its maximu intensity in the lower right quadrant of the vi ble surface. All maria seem to be related. If th force was weaker near the left limb, upheava and subsequent sinkings would have produce here lesser modifications of the original surfa than they did in the Imbrian area. The hypothes of a trigger impact is less satisfactory in the low left hand quadrant than it is for other parts the surface.

Mare Serenitatis and Mare Tranquillita (Plates 10–11—10–16). These seas are larg than any of the preceding ones. Mare Serenita is isolated except for three gaps, one on the sou to Tranquillitatis, one on the right to Imbriu and a small irregular one on the north to an e tension of Mare Frigoris. Plates 10–13 and 10– exhibit several interesting features of the le hand floor of Serenitatis. One of these is a fi continental shelf. Farther in on the floor are th great north-south ridges that merge at the northern and southern ends and which are a co tinuation of the ridge system of the right han portion of Tranquillitatis, Plate 10–17. Plate 1 14 shows a beautiful row of half-craters along th southern shore extending left from Menelau Along the left shore, Le Monnier has lost all of i seaward wall to become a bay. Just to the north Le Monnier is Posidonius, god of the seas, fitting represented by one of the greatest of the ring plains. Its right wall is almost broken at one poi and it duplicates almost perfectly the situatic of Gassendi at Humorum and Pitatus at Nubiu Few ghosts are found on the floor of Serenitati

A remarkable feature of Mare Serenitatis the very dark edging just inside the shore lin well shown by Plates 10–14, 3–13, 3–14. The ed

Plate 10–13. Mt. Wilson and Palomar Observatories, 1957 May 6ᵈ 04ʰ 49ᵐ UT, phase 6.20 days, colongitude 351°. Sunrise terminator on Mare Serenitatis.

Plate 10–14. Mt. Wilson and Palomar Observatories, 1954 July 20ᵈ 09ʰ 31ᵐ UT, phase 19.88 days, colongitude 148°. Sunset on left shore of Mare Serenitatis.

Plate 10–15. Mt. Wilson and Palomar Observatories, 1958 July 7^d 10^h 32^m UT, phase 20.11 days, colongitude 161°. Right hand part of Mare Serenitatis and the crater pit on Linné.

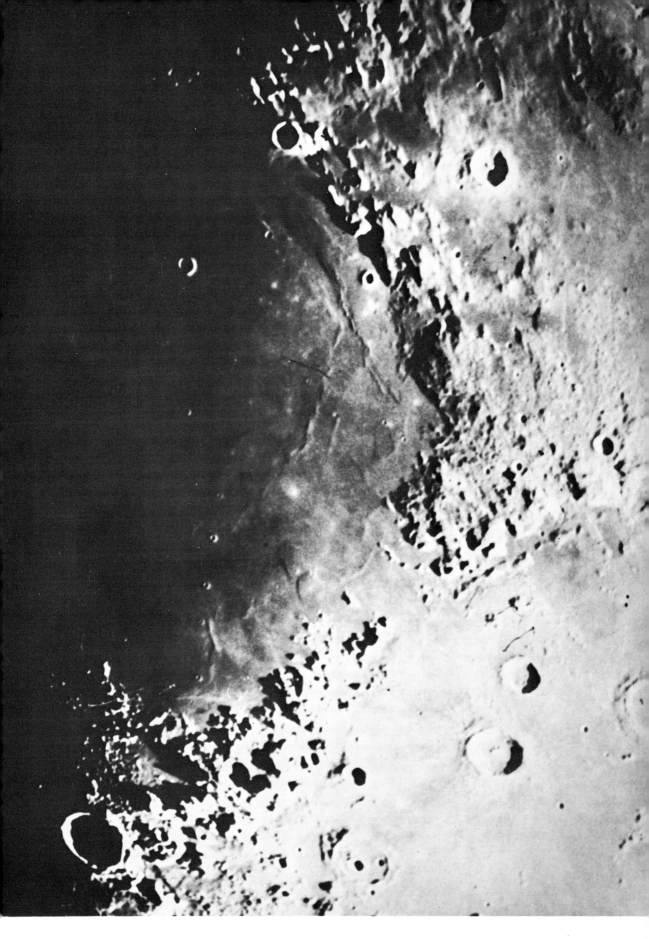

Plate 10–16. Mt. Wilson and Palomar Observatories, made one minute after Plate 10–15, using Kodak II–0 plate sensitive to blue-green and shorter wavelengths. Covers same area as Plate 10–15.

ing extends through the gap into Tranquillitatis, where in many places the floor is of the same dark shade. From the analogy of terrestrial lava flows, one might guess that the edging is a result of oozing of molten rock after the main process of evolution had been completed. The idea has been advanced, because of the extension to the south, that the edging indicates a flow of lava from Tranquillitatis into Serenitatis. Although it is possible that such a flow did occur, it must have been independent of the edging, as is indicated by the presence of the edging along both right and left hand shores.

On the righthand floor of Serenitatis is Linné, the most controversial of all the lunar craters. It appears as a tiny hole in a white patch in Plate 10–15, but the hole seems obscured in Plate 10–16, made with short wavelengths. Whether craters sometimes emit residual gases or not is discussed in Chapter 16. In the closeup of Linné

in Plate 10–16–A, it looks like a normal crat

Tranquillitatis (Plate 3–10) is a large ma with a very irregular and generally low sho line. To its left is a large, partly blocked chan which connects it to Mare Foecunditatis. To t south a strait connects it to Mare Nectaris. Rig of it is the jumbled, partly marshy, area north Ptolemaeus. The right hand part of the flo (Plate 10–17) shows a section of perhaps t finest ridge system on the whole moon. It exten from Theophilus north through Tranquillita to connect with the system in Serenitatis and th form one of the most spectacular features of t moon. This closely meridional ridge system, mo than 800 miles long, and another system whi almost duplicates it in Oceanus Procellaru surely cannot be the results of localized impa forces.

Mare Imbrium (Plates 10–18—10–24). E cept for Oceanus Procellarum, Mare Imbriu

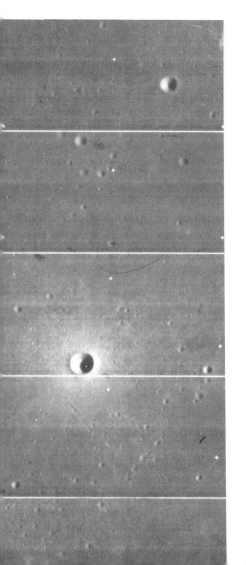

Plate 10–16–A (Far Lef Lunar Orbiter IV photogra of Linné crater from an a tude of only 2705 km, the b view of the crater ever mac Here it appears like a typi small crater, although obse ers have reported changes its pit and shape and obscu ation of its pit.

Plate 10–17. Mt. Wilson an Palomar Observatories, 195 July 20^d 09^h 31^m UT, pha 19.88 days, colongitude 148 Kodak 33 plate, no filte Ridge system north fro Theophilus, through Ma Tranquillitatis into Mare S renitatis.

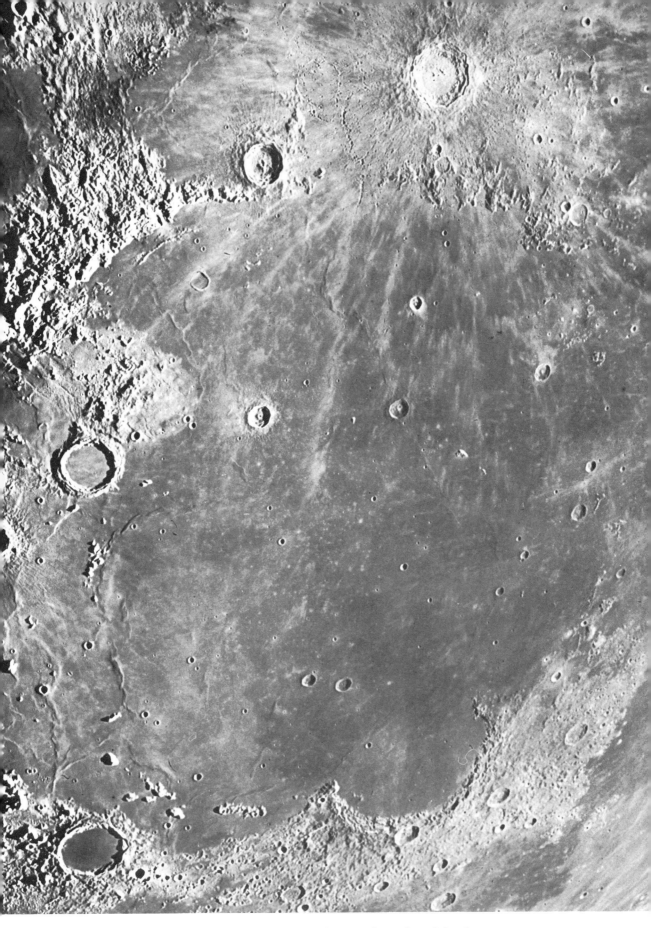

Plate 10–18. Sunset at left shore of Mare Serenitatis. See Plate 3–2 for technical details.

Plate 10–19. Portion
Plate 10–18. Detail of 1
floor of Mare Imbrium.

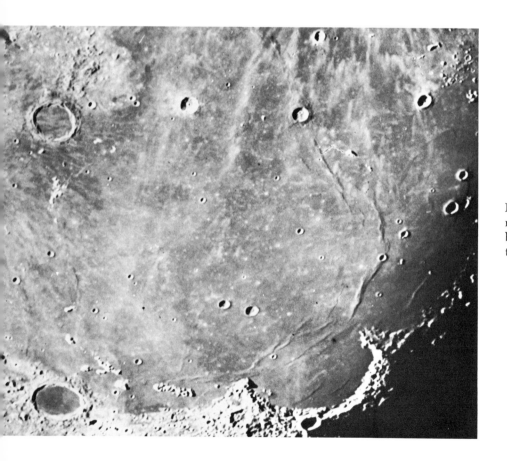

Plate 10–20. The ridges
right hand part of Mare I
brium. See Plate 4–10 f
technical details.

Plate 10–18) is the largest of the maria. Plates 10–19 and 10–20 show the great central ridge ring of Mare Imbrium, a roughly circular low ridge around the center of the mare. At places it is so low that it can be observed only for a short time after sunrise and before sunset. The diameter is great enough that the right and left sides must be observed at different phases. On the south it includes Timocharis and Lambert. The rays of the former prove that crater to be younger than the mare. Plate 10–20 shows this area and the right side of the ring. Plate 10–21 is an enlargement of the region near Lambert and shows that crater in the lower right. Just before reaching Lambert the ridge is joined by the principal northern rays from Copernicus. Often the statement is made that rays are not elevated. The statement perhaps is true of most but not of all rays: Plate 10–21 shows that these Copernican rays lie along a ridge. From Lambert the ridge continues to Caroline Herschel and then turns to the lower left to block the entrance to Sinus Iridum. Next it touches the Straight Range, then continues left to include the Teneriffe Mountains (Plate 10–19) and the separate mountain Pico. At Pico it turns southward once more to Spitzbergen, a small mountain range north of Archimedes, then extends to the upper right toward Timocharis. This is a complex ridge, and not the boundary of a sunken area such as may be found, for example, in Mare Crisium and in Sinus Aestuum.

Perhaps a third of Imbrium lies within the ridge system. This central area is very different from the remainder of Imbrium and from the other maria. The ring in most places has the general appearance of being an outward, compression wave of molten rock from some center of disturbance. If in any lunar area there has been asteroidal bombardment amounting to more than triggering action in the formation of a mare, it has been here.

Plate 10–22 shows that a useful picture need not be a pretty one. The highlights are so dense

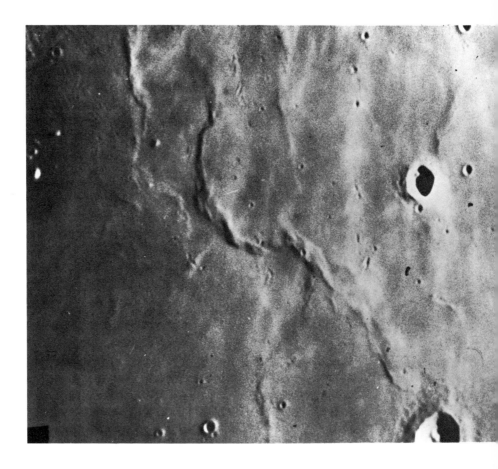

Plate 10–21. Mt. Wilson and Palomar Observatories, 1955 October 9^d 11^h 25^m UT, phase 23.21 days, colongitude 188°. Detail of ridges in right hand part of Mare Imbrium.

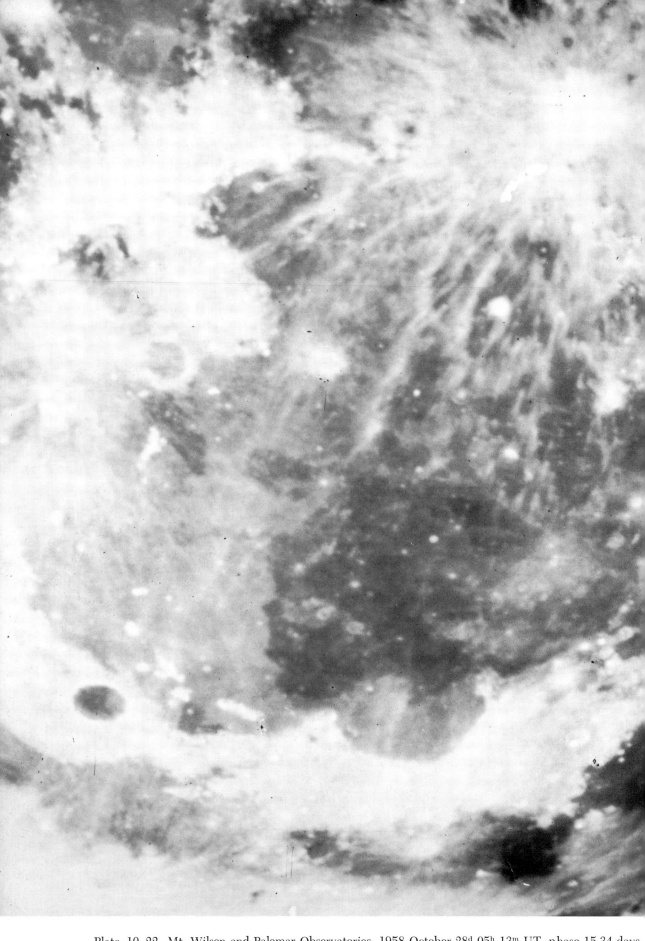

Plate 10–22. Mt. Wilson and Palomar Observatories, 1958 October 28^d 05^h 13^m UT, phase 15.34 days, colongitude 97°. Kodak IV–N plate with infrared filter. The asteroidal impact area.

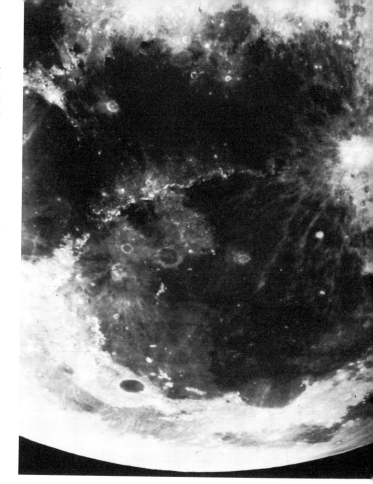

Plate 10–23. Mt. Wilson and Palomar Observatories, 1958 August 1^d 08^h 16^m UT, phase 15.57 days, colongitude 105°. Kodak IV–N plate. This plate was printed longer than Plate 10–22 to show detail in brightest regions.

n the negative that they are merely white paper n the print. The plate used was a Kodak IV-N nfrared, which is supposed to be too contrasty o have any value in lunar photography. It was xposed only a half-day after the full moon phase. Iowever, the high contrast brings out certain ypes of detail on the dark mare floor better than ny other plate available for this area. As a result, everal hours have been spent in almost memoizing the story it tells. The greatest difference etween it and other pictures is the clarity with hich it exhibits a large triangular area within e central ring, showing it to be distinctly darker an the remainder of the floor. This is the aproximate area pinpointed by several astronoers as the postulated site of an asteroidal imact, which they suggest was the cause of the are. We can be quite certain that an asteroid, erhaps one belonging to the earth-moon sys-m, did strike in this place.

Before we fully accept this hypothesis of the igin of Mare Imbrium we must look at other atures of the picture. To the south of the imct area Copernican rays are seen traversing veral hundred miles of the mare floor. They op at the southern boundary of this darkest ea. When Copernicus came into existence the uthern part of the floor could not have been tremely hot, for if it had been it would not exbit the rays. Certainly the molten rock must ve been hot enough originally to prevent the rmation of such rays. It seems quite certain, erefore, that the southern part of the floor of hbrium is considerably older than Copernicus. ere are two possible explanations for the stopge of the rays at the dark area. The first is that e outer mare area had cooled very much bere Copernicus was formed but that the actual ea of the earlier impact still remained very hot. lis is certainly a reasonable explanation. The

Plate 10–24. Correction of foreshortening at Sinus Iridum (see Appendix).

other is that the impact occurred after there was much cooling of the mare surface and that the impact, instead of creating the mare, merely produced profound modifications within it. Which of these is true must be decided from a study of the whole lunar surface.

Other interesting features exist, some of them best shown by underexposing a photograph of the full moon, as was done for Plate 10–23. In it Mare Frigoris is seen as a long series of small connecting seas, separated from Imbrium by the Caucasian-Alpine-Iridum island. Frigoris parallels the northern shore of Imbrium and appears to have been formed at the same epoch as that larger sea. The long island was an area of resistance. Another such area of resistance is the partially sunken area extending from the Apennine scarp to Archimedes (Plates 10–18, 10–22, 10–23). On photographs of the full moon a glance at Archimedes should serve as a warning against accepting any one observation in deciding the nature of a lunar feature. In these pictures Archimedes appears definitely to have a broken left wall. Pictures made when the shadows are longer, as for example in Plate 10–19, show that no indication of any such break exists. The one phase emphasizes differences of shade of the surface; the other exaggerates the effects of differences of elevation.

Aristillus and Timocharis appear under a high sun (Plates 10–22 and 10–23) to consist mainly of a ring of bright dots with another such dot at the center of the ring. Under a lower lighting they are shown to be typically fine examples of explosive craters, and ringed plains. Study of them at different phases proves the bright spots to be craters or in some cases half-craters around the rims.

Outside the central ring Mare Imbrium follows the same evolutionary pattern as do the other maria. To the left is the Apennine scarp, far the grandest to be found. The scarps continue northward as the right boundary of the Caucasus Mountains and to the right as the northern boundary of the Carpathians. A broad peninsula to Archimedes (Plates 10–18, 10–22, and 10–23) is, apparently, composed of partially sunken fragments of the original Apennines. In the neighboring area there is much "marshy" ground. Inside this left hand portion are found detached islands of resistance. Mount Pico is a remnant of

an old and very large craterlike ring. Everywh[ere] in its left hand area Imbrium appears to foll[ow] the usual pattern of mare evolution and sho[ws] nothing of a nature different from the oth[er] maria.

The northern shore of Imbrium (Plates [10–] 18, 10–22, and 10–23) reveals a less precipit[ous] scarp than the Apennine one. On this northe[rn] shore is Sinus Iridum, the Bay of Rainbows. So[me] able advocates of the impact formation of [the] mare have postulated that this is the site of [the] initial impact of the asteroid. But when fo[re]shortening is corrected (Plate 10–24) by pho[to]graphing one of the Wright globes, the bay [is] seen to be deep rather than shallow, its boun[d]ary derived from two sides, and parts of t[wo] others, of an old hexagonal mountain-wall[ed] plain. The history was almost the same as th[at] of the Straight Wall (Plates 10–1—10–3).

Sinus Aestuum is a large bay just south [of] Eratosthenes. On Plate 10–18 its floor exhib[its] one of the finest sunken areas to be found on t[he] moon.

What, in sum, is the origin of the maria? [To] the facts which have been presented in this cha[p]ter and to many more like them, which have be[en] omitted because of lack of space, must be add[ed] special studies of peculiar areas. The evide[nce] of the area to the south of Mare Humorum, w[ith] its mare-like features that have almost obliterat[ed] Tycho's major rays through an arc of fully 120[°;] the wild marshy area south from Serenita[tis,] right from Tranquillitatis and extending alm[ost] to Ptolemaeus; the long, complex Mare Frigo[ris] and the general mare areas—all seem to dema[nd] hypotheses along these lines:

1. The maria and their plains have result[ed] from a great, convulsive force acting on the s[ur]face of the moon, followed by sinkings and t[he] outflow of molten rock.

2. Meteorites and asteroids striking in [un]stable areas may have had a trigger effect, in[iti]ating the process described in the preceding pa[ra]graph, but they cannot by themselves have p[ro]duced the maria in the form in which they n[ow] exist.

3. The present form of that part of M[are] Imbrium which is within its central ring is [an] exception to the foregoing hypothesis.

4. The maria, with some possible exceptio[ns,] are younger than the mountainous areas.

1 THE NATURE OF THE TYPICAL

MOUNTAIN-WALLED PLAINS

or more than a century the various craterlike ormations of the moon have been classified as ountain-walled plains, ringed plains, crater ngs, craters, craterlets, and crater pits. Probably better classification could be devised today, but ew nomenclature would only result in confu- on; it seems best to wait until the present dis- greement about the nature of these objects has een resolved before changes are made. The aditional names are descriptive and have the erit of implying nothing with regard to origin, bout which there is a great deal of speculation. ne of the new terms that is sometimes used and suggestive of type of origin is "endocrater," hich would apply to any craterlike formation onsidered in the context to be the result of uses that were internal—within the moon itself. ven a shallow, local, accidental depression ould then be considered as an endocrater.

Conspicuous among those considered here be endocraters are the mountain-walled plains. eison describes them as follows:

"Walled plains extend from 40 to 150 miles in diameter, and are seldom surrounded by a single wall, but usually by an intricate system of mountain ranges, separated by valleys, crossed by ravines, and united to one another at various points by cross-walls and buttresses; all usually, however, subordinate to one or two principal ranges, forming a massive crest to the rest. To- ward the exterior and interior extend numerous projections and arms, at times rising even above the wall and at others low, short, and insignifi- cant. Occasionally, as on Schiller and Posidonius, these arms extend throughout the greater por- tion of the interior, or even divide it into two portions. Toward the exterior, these branching arms and projecting buttresses occasionally unite two or more walled-plains together, and at times these rise into considerable ridges, often enclos- ing long valleys. The interiors of the walled- plains are as a rule comparatively level, some- times, as in Plato and Archimedes, only broken by a few mounds, or perhaps by a crater cone or so; but more usually the interior is interrupted by a number of small irregularities, as ridges, mounds, or crater pits, as in Maginus and Ptolemy; whilst at times these irregularities assume considerable dimensions, as in Posidonius, Gassendi, and Catharina. Though many are roughly circular in shape, others possess very irregular outlines, appearing more like several confluent plains, or like a space enclosed by intersecting mountain chains rather than as true independent formations.

"Though commonly classed under the crateriform formations of the moon, the true walled-plains would appear to be related rather to the mares or plains, more especially to those mares bordered by great highlands and mountains like the Mares Crisium and Serenitatis, to which certain of the great walled-plains, as Clavius, Maginus, Ptolemaeus, Hipparchus, and Schickhardt, bear a considerable resemblance, though on a smaller scale—a circumstance that did not escape Maedler. A close examination of such examples of the walled-plain as these would suggest their being low-lying bright plains surrounded by mountain ranges and extensive highlands, rather than actual independent formations bearing any relation to true volcanoes, and, as Maedler remarks, had Clavius possessed a dark interior, and been nearer the centre, Riccioli would have probably classed it as a mare, and the same holds good with some of the others."

There are certain data which must be explained, or at the very least not contradicted, by any satisfactory hypothesis concerning the nature of the typical mountain-walled plains. These are:

1. They are very shallow in respect to their diameters.

2. They have low or no external walls.

3. They seldom have central peaks.

4. The walls are not circular, but rather are rectangular, square, and especially hexagonal.

5. Many small craters are found on their rims.

6. The floors tend toward smoothness.

7. They become inconspicuous under a high sun.

8. They have diameters from approximately 50 to 150 miles. The lower limit is inherited, and arbitrary.

9. They are found only in mountainous areas.

10. They resemble those of the maria which are bounded by scarps.

11. Scarps and lines of small craters sometimes are observed along lines which are the mu-

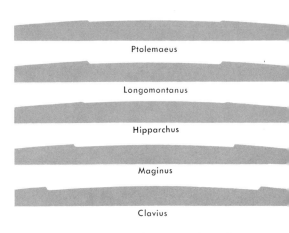

Figure 11. Vertical sections through typical walled plains.

tual extensions of the boundaries of neighboring plains.

12. The walls are not continuous; gaps are common.

13. They have no ray systems.

14. Some—Ptolemaeus (Plate 11–2) is the best example—have numerous "ghosts" on their floors.

15. In general, if a piece of paper is laid over a photograph so that it just touches the inner edge of the boundary scarp, there is nothing visible to indicate that any unusual depression has been covered.

The accompanying diagram attempts to show vertical sections of several of the largest walled plains on a true scale. Heights of the ridges are not known accurately and it is probable that the average heights are even less than those used in the diagram. This statement applies especially to Longomontanus, for which the old measurements appear to be clearly excessive. Usually the wall contains some mountains much higher than the average. Often gaps occur in the walls. Plate 9–1, showing Ptolemaeus at sunset, exhibits both of these features.

Some lunar features conform to most of the characteristics of mountain-walled plains but differ more or less widely from the typical examples. In some cases the differences are so great that one may wonder whether the object should not be more properly classified as a ringed plain. Perhaps a category for this kind of hybrid should be created, intermediate to the classical divisions. Such intermediate examples show evidence

11–1. Mt. Wilson and Palomar Observatories,
May 24ᵈ 11ʰ 36ᵐ UT, phase 21.63 days, colongi-
.72°. Kodak IIa–O plate, no filter. The Ptole-
area near sunset.

Plate 11–2. Mt. Wilson and Palomar Observatories,
1956 October 26ᵈ 12ʰ 56ᵐ UT, phase 22.35 days,
colongitude 177°. Detail of floor of Ptolemaeus.

more or less explosive action and they exhibit
ore than traces of external walls.

The trio seen on Plates 3–19 and 11–1, Ptole-
aeus, Alphonsus, and Arzachel, have all been
nsidered to be mountain-walled plains. Ptole-
aeus is a typical one. Alphonsus is definitely
lated to it generically and does exhibit the
xagonal wall. However, there is also evidence
explosive action in the formation of Alphonsus,
d there is a noticeable external wall. It would
pear to be probable that the explosive action
me late in its formation. Arzachel appears also
have been initiated by collapse, but the later
plosive action was so violent that there could
no objection to classifying it as an explosive
nged plain.

In judging the existence of an external wall,
re must be exercised lest a contiguous feature
ove deceptive. If a mare is nearby, for instance,

the slope to its surface may create a false impression of a wall. A large depressed area, such
as the one bounded by Arzachel, Alphonsus,
Ptolemaeus, Albategnius, and smaller features to
the south on Plate 3–19 may produce such an
illusion. Plato on Plate 10–19 is listed as a walled
plain by both Neison and Goodacre. Possibly it
is, but if so it was much altered by the formation
of Mare Imbrium. It is located on the narrow island of ground that separates Maria Imbrium
and Serenitatis from Frigoris. The slopes to these
maria, which were formed later, may possibly
have produced a fictitious external wall. Archimedes, on the same plate, is a ringed plain, and
is so listed by Goodacre despite Neison's erro-
neous classification.

In contrast to those craterlike formations
that have considerable external slopes, there is
little tendency for the typical walled plains to **109**

be circular. By way of example, Plates 11–1, and 11–2 and 3–19 show Ptolemaeus as an almost perfect hexagon. Albategnius, another hexagon, lies to the left of it. Hipparchus, just north of Albategnius, has almost lost its lower right wall, but the remaining southern boundary of the plain is more nearly part of a square than anything else. The boundary wall of Hipparchus is nearly as low as is that of Deslandres, and even while the sun still is moderately low (Plate 4–7) it becomes difficult to find. Directly right of the southern side of Ptolemaeus is a small unnamed plain (Plate 11–1), which is nearly a perfect rectangle.

Farther south (Plates 3–19 and 8–1) is the sequence Purbach, Regiomontanus, and Walter. Regiomontanus is a rather good square, except for the fact that Purbach has cut off its lower right hand corner. Walter also shows somewhat the same tendency. To the right of these is the

Plate 11–3. Clavius corrected for foreshortening (see Appendix).

very large and very shallow old plain, Deslandre It also roughly approximates a square. It appear that the largest of the mountain-walled plain are the oldest and perhaps the most typical one Deslandres is so old that several later feature have formed on it. It also is so shallow that was not listed by the early observers. Howeve when the sun is very low (Plates 13–1 and 13–2 it is revealed as one of the finest lunar depres sions. Similarly Janssen (Plates 15–2 and 15–5) easy to miss, except when the sun is quite low, be cause of both its low walls and the many late features that camouflage it.

Clavius, the second largest of the typica walled plains, is found near the top of Plate 8– Almost at a glance one observes the many crate at and near the rim, the lack of external wall, th general smoothness of the floor, and the beautifu arc of craters on that floor, starting with Ruthe ford on the southern wall. The true form of Cla vius is difficult to distinguish because it is nea the southern limb of the moon. However, th floor is observable as a somewhat irregular poly gon, marred by the more recent Rutherford one corner. This is revealed clearly by Plate 11– in which the area near Clavius is corrected f foreshortening; see page 198 of the appendi Close examination reveals that there is somethin peculiar about the southern side. The souther part of the floor is seen to extend southward at higher level, as a pass between extensions of th walls. The right extension becomes an outer le wall for Blancanus and the left extension serv as an outer right hand wall for Gruemberge Each of these is a walled plain more than 50 mil in diameter. This pass is probably one example very old surface features that have become a most fully camouflaged. It is related to a muc larger and very old depression just to the left Clavius. Clavius apparently was formed in right hand section. When the lighting is suf ciently low the wall of the large depression ca be glimpsed running to the upper left from point just north of Clavius. As seen in Plate 8– the depression next makes a turn to the uppe right and passes barely south of Gruemberge This slightly depressed area may be the olde feature observable on the moon today.

Grimaldi, near the right limb, is one of th largest of the walled plains. Foreshortening di guises its form, but the globe projection metho

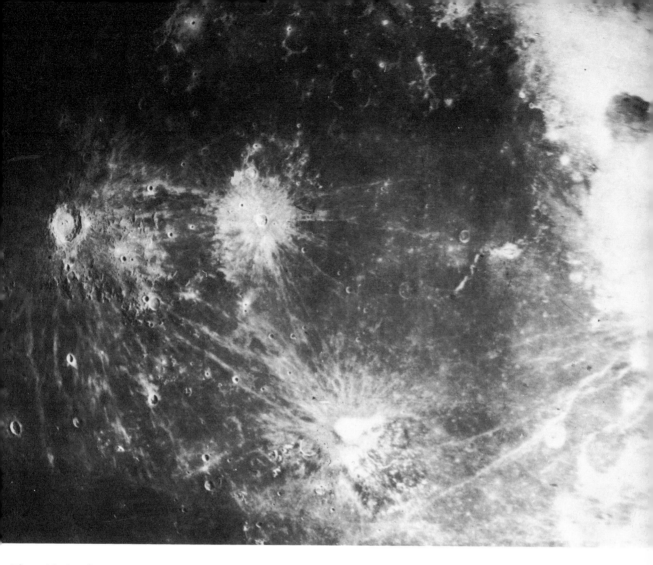

Plate 11–4. The Copernicus-Kepler-Olbers-Seleucus-Aristarchus oversystem
of rays, partially corrected for foreshortening (see Appendix).

Plate 11–4) shows it in nearly its true shape.
is revealed as a hexagon with a mutilated
rthern wall where a pass leads to Oceanus
ocellarum. Grimaldi is morphologically truly
connecting link between the maria and the typi-
l walled plains. On its shoreline on the right
n be seen the remains of several craters whose
award" walls sank during the formation of
rimaldi.

If the walled plains were due primarily to
plosions, either internal or external, we would
t expect to find these polygons. But we might
ll expect them if the typical walled plains
re in fact somewhat similar to what geologists
ll calderas: craters formed by collapse, often
ong a fault line. (The walled plain Schiller re-

sembles a terrestrial "graben," a sunken area between faults.)

The evidence favoring a caldera-like explanation of the walled plains is emphasized by a study
of the lines along which the sides lie. In Plate
11–1 examination of the northern wall of Ptolemaeus shows that it contains a line of small
craters. As the line of craters is extended to the
left from Ptolemaeus it merges with the northern
wall of Albategnius. The coincidence is too great
to accept as accidental. This is perhaps the most
striking example of this kind of linear relationship between large neighboring walled plains
that are not in actual contact.

As another example, the left wall of Walter
can be observed, on Plate 8–1, continuing south-

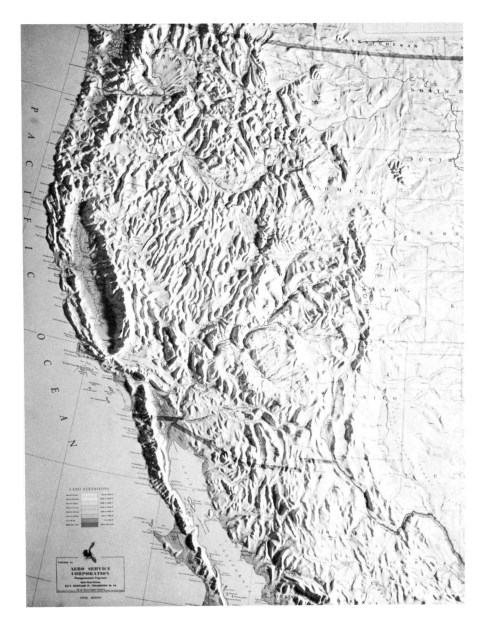

Plate 11–5. The great mountain-walled plain in Califonia. (Courtesy of Aero Scienc Corporation.)

ward in a conspicuous manner beyond the floor for a distance equal to the diameter of the plain, and can be picked up at intervals for a much greater distance southward. Under the morning lighting of this plate Walter appears like the heel of a giant's footprint.

Plate 8–5 shows a large area of the rough section of the moon south of Mare Nubium. In it there are several places where the rim of a craterlike formation continues conspicuously beyond the depressed area, either as a scarp or as a row of craters.

The small craters at or close to the rims these plains are too numerous, as compared wi the numbers in the surrounding areas, to be a cidental. This fact may be easily observed examination of Plate 8–5. Clavius, the secor largest of the walled plains, exhibits these sma craters. To the right of Clavius, Scheiner has nice ring of them. Below and slightly to the le of Clavius, Maginus shows perhaps more of the than any other plain. Directly across from Mag nus, Longomontanus shows a similar formatio North of Longomontanus, Wilhelm I has a rath elicate ring of small craters. Near the bottom the picture, the partially ruined Pitatus con- nues the story. This excess of small craters quite efinitely indicates weaknesses in the rock at ese places. In many cases lines of these craters ppear to resemble somewhat our terrestrial owhole craters. Observed strings of them ubtless lie along fractures of the rock.

These data indicate strongly that typical ountain-walled plains, examples of which are sted below, are not due primarily to explosions ther from internal or from external causes. The umbers following the names are the (x, y) co- rdinates from the map of Plates 3–1 and 3–2.

lbategnius (66, 84)	Maginus (72, 119)
lavius (75, 124)	Ptolemaeus (74, 82)
eslandres (74, 106)	Purbach (72, 99)
ipparchus (65, 78)	Scheiner (80, 124)
nssen (38, 114)	Walter (70, 106)
ongomontanus (84, 118)	Wilhelm I (85, 114)

In searching for the cause of these objects ne may be tempted to accept the hypothesis of he collapse of great domes. Small domes, and raterlets that have resulted from the collapse f large domes, have definitely been found on he moon. Great domes have been postulated on he earth. Their terrestrial existence, however, s controversial, except perhaps for a few in- tances. For example, the central plain of Cali- ornia has been cited as being the result of dome ollapse, but instead almost surely it was caused y the tipping of the Sierra block. This great errestrial "mountain-walled plain" is shown by late 11–5, in which the height of the boundary vall is exaggerated. If large domes ever existed n the moon they were formed by great upward luid pressure beneath the surface. Such a pres- ure might have been gaseous in nature, or the esult of steam from rock-locked water in the nterior. A more likely guess as to the agent vould be expansion and fluid rock resulting rom changes of phase of some of the minerals n the lunar mantle. The collapse of such domes ould be due to fracturing of the surface rock s they were lifted too high, or to triggering im- acts by meteorites. Meteoritic impact cannot, owever, account for the form in which we now ee the great mountain-walled plains. At present he dome hypothesis of the origin of these ob- ects cannot be entirely rejected, but it does seem

improbable. The areas appear to be too small for lack of isostasy to serve as a satisfactory hypothesis for their origin (although this may be possible in the case of the largest maria). Perhaps, after all, not all these features had the same kind of origin. Whereas some of them may actually be related to the maria genetically, others may merely have a superficial morphologi- cal resemblance. Certainly the origins of the craterlike formations (ringed plains) that exhibit considerable external slopes must be different from those of the mountain-walled plains. For the former some type or types of explosions are indicated.

A last word on the caldera hypothesis: it is of course dangerous to argue from morphological resemblances between features on the moon and on the earth, which is much greater in mass and possesses an atmosphere. It would be equally wrong to neglect such features completely. Plate 11–6 is an aerial photograph of an active caldera in the Galapagos Islands. The picture was made by the U.S. Air Force in cooperation with the Ecuadorian government and supplied through the courtesy of Dr. Jack Green, a geologist who is making a special study of calderas.

The resemblance of such terrestrial calderas to mountain-walled plains such as Ptolemaeus, Grimaldi, Clavius, and the large old plain that includes the Straight Wall must not be ignored. The same holds for ringed plains such as Era- tosthenes, Theophilus, and Archimedes, and for the Maria Crisium and Serenitatis, even including their continental shelves.

There is need for a great deal more research. At this stage these judgments concerning moun- tain-walled plains appear to be justified:

1. Some of them are closely related in nature to the maria.

2. If any explosions occurred in their forma- tion, they were at most merely triggerlike ones.

3. They are large sunken surfaces in moun- tainous areas of the moon, and are usually of polygonal form.

4. They often show specific relationships to neighboring features.

5. Small craters at or near their rims indicate that there has been sinking along fault lines.

6. Some of them appear to be the oldest fea- tures observable on the moon today.

7. Probably most are some form of caldera. 113

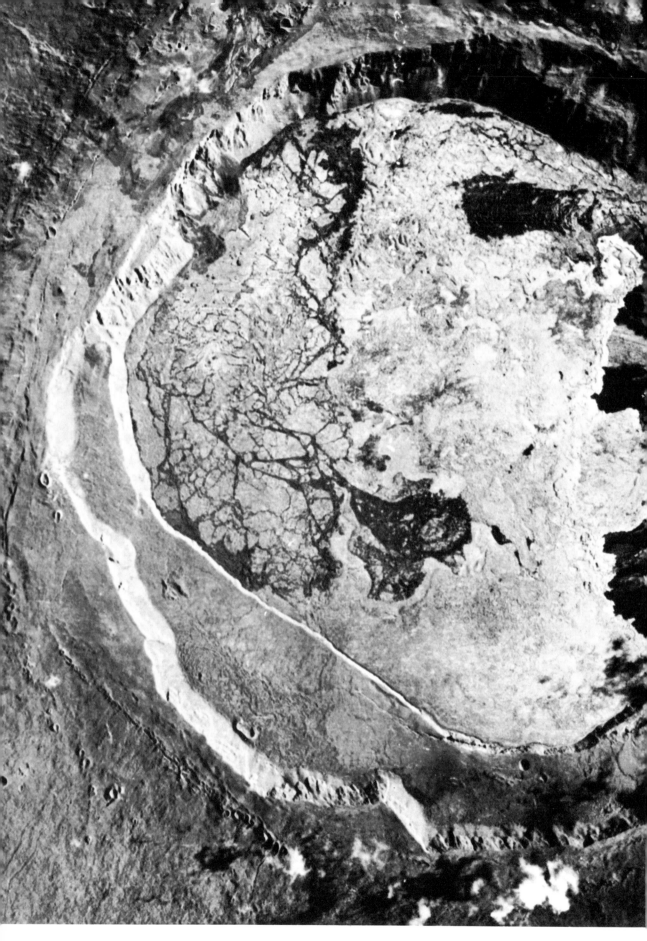

Plate 11–6. Active caldera in the Galapagos. (U.S. Air Force
in cooperation with the Ecuadorian Government.)

2 THE EXPLOSION CRATERS

he evidence offered by the surface of the moon ndicates that in general its evolution was comparatively quiet; most of the surface features are imply not of a kind that indicates very violent explosions. The maria appear largely to have een the results of subsidence and outflow; there little or no reason to believe that explosions ere a significant factor in their evolution except the central part of Mare Imbrium, Plate 10–22. he same appears to be true of the next largest ass of features, the typical mountain-walled lains, which have maximum diameters exceeding 150 miles.

However, large and violent explosions have aken place. The two best known of these are iscussed descriptively in Chapters 7 and 8. At ll moon phase a person viewing the scene can lmost imagine the explosions to be occurring ow. Eleven of the principal examples of such iolent explosions are marked on the accompanying plate, keyed by number to the list below it. he numbers in parentheses are the brightnesses s stated by Neison; see Chapter 2.

Any satisfactory hypothesis concerning explosion craters (both the ringed plains and the maller craters) must explain, or at least not contradict, the following facts:

1. The explosion craters have definite external walls.

2. Many of the explosion craters brighten much more than do the neighboring features under a high sun.

3. Others, such as Eratosthenes (Plates 7–1—7–5) almost disappear when shadows are short.

4. The inner walls of some of the larger craters, such as Copernicus, have terraces.

5. Craterlets are commonly found around the rims of all except the smallest craters.

6. Unlike the typical mountain-walled plains, which are found only in mountainous areas, explosion craters exist in all kinds of terrain.

7. Lines of craterlets exist on the outer walls and in the neighboring areas of some of the explosion craters. This is much more conspicuous for Copernicus and Tycho than for any of the less violent ringed plains. Sometimes these craterlets coalesce to form rills (clefts); see Plates 7–1, 7–2, and 8–7. Usually the rills tend to be radial to the primary explosion but some rills point in other directions or even follow curved loci.

8. Some of the outer slopes of explosion craters exhibit valleys that give at least a casual impression of having been formed by erosion. Aristillus (Plate 10–19) is an excellent example.

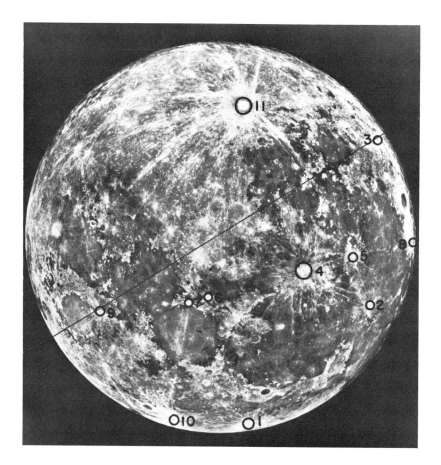

Plate 12–1. Ringed plains of apparently unusually violent origins. See Plate 5–2 for technical details.

1. Anaxagoras (7)
2. Aristarchus (8–10)
3. Byrgius (6–7, crater 8)
4. Copernicus (9)
5. Kepler (7)
6. Manilius (8)
7. Menelaus (8–9)
8. Olbers (6)
9. Proclus (8–9)
10. Thales (6–8)
11. Tycho (8 crest)

9. Central mountains or groups of hills are common within these craters, whereas they are rather rare within the typical mountain-walled plains.

10. Most, although not all, explosion craters exhibit a tendency toward circularity. This contrasts strongly with the rims of the mountain-walled plains which show a definite tendency toward polygonal shape.

11. Perhaps each of those craters which brighten definitely under a high sun shows at least one ray (bright streak) which belongs to it. In some cases such as Tycho, Copernicus, and the starlike craterlets on the outer slopes of Furnerius and Stevinus, the ray systems are large and complicated. It is possible that systems of faint rays belong to some craters which do not brighten much. Mountain-walled plains do not have ray systems.

Some craterlets, such as those on the rims to the right of the walled plains Furnerius and Stevinus (Plates 3–4 and 4–9), also indicate tremendous, although highly localized, explo-

sions. Near full moon the bright craters and their rays are as conspicuous as are the maria and the great mountainous areas. The violence of their explosions is attested to not only by the changed albedo over large areas but also by the ray systems and in some cases by the numerous craterlets in the neighborhood. Still, the total area of all these violent features is only a small fraction of even the mountainous areas in general.

The ray systems prove that at least some of these explosions were among comparatively recent catastrophic events. Nevertheless, the mare-like areas right of Tycho cannot have subsided much earlier than the time that Tycho (Plate 10–4) was formed. If they were formed before Tycho they had not yet cooled sufficiently to permit Tycho's major rays to cross the area. Tycho's lower right rim (Plate 8–6) is slightly cut by an oval partial depression, which must have been formed more recently than Tycho. Palus Somnii to the right of Proclus (Plate 10–12) cannot have antedated Proclus, although the bsidence may have occurred at the same time the explosion.

The evidence of extreme violence in all ese cases is so great that one must agree with lbert, Baldwin, Urey, Kuiper, and others that, least initially, these craters were formed by eteoritic or asteroidal impacts. No other hypothesis seems adequate to account for the tremendous release of energy on the rather quiet oon. The small, starlike craterlets, which dot e surface of the full moon, may be assumed be entirely the results of meteoritic impact.

But at least in the cases of Tycho and Copernicus, the numerous craters that have formed their vicinities testify to the probability that bsequent volcanic action took place on a large ale. This evidence is strengthened by the easily servable terraces on their inner walls, which dicate subsequent collapsing that increased the aters' diameters. The hypothesis of subsequent olcanic action is almost demanded, moreover, y the presence of craterlets on the rims, indiiting faults along which subsidence occurred.

The arc of craterlets on the left hand floor of Tycho surely must be volcanic. This volcanism may be merely an indirect result of the impact explosion, but it does exist. Tycho, Copernicus, and probably other large craters of this class, then, must be of a composite nature: impact-volcanic.

In contrast to the very bright craters, which clearly are, at least in part, of impact origin, there is a second class of large explosion craters that show little or no difference either of shade or color from the surrounding areas, regardless of the height of the sun. The external walls indicate that explosions did occur, but it appears probable that the explosions may have been much less violent than those connected with impact craters of equal size. In general these second-rank explosion craters do not exhibit ray systems or an unusual density of nearby craterlets. Eratosthenes, which may be an immense mountain with a large crater at its summit, is perhaps the best example. Quite possibly the top of the mountain collapsed into a partially hol-

ate 12–2. Large ringed ains of apparently moder-ely violent origin. See Plate -2 for technical details.

. Aliacensis

. Aristillus (4–4½)

. Arzachel (4)

. Atlas (5)

. Autolycus (3–5)

. Cassini (4)

. Cyrillus (4, crater 7)

. Eratosthenes (3½–4)

. Hercules (5, crater 8–9)

. Lambert (4)

. Theophilus (5)

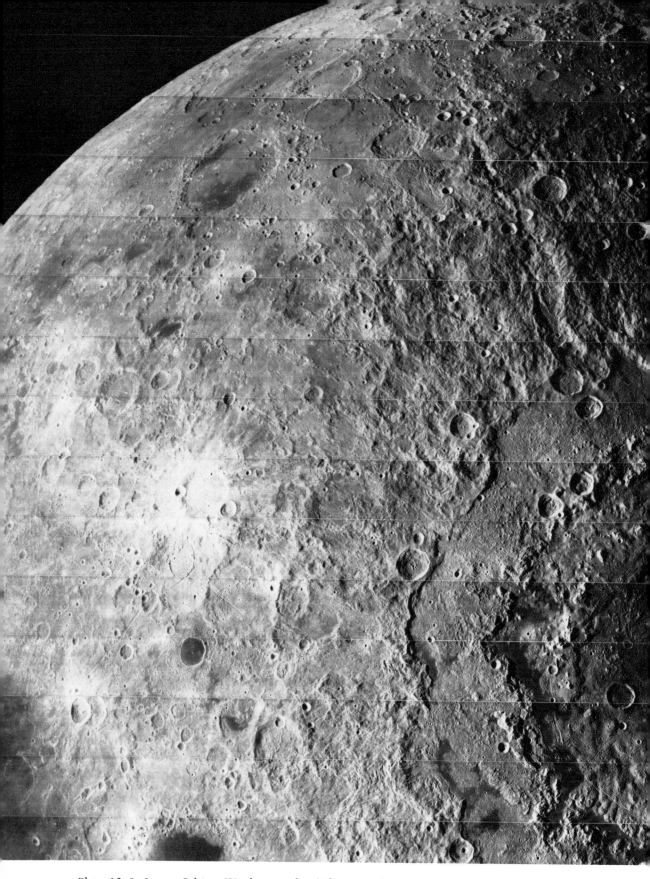

Plate 12–3. Lunar Orbiter IV photograph of the great formation named Mare Orientale on the lunar farside. An asteroid may have triggered vast endolunar forces in the creation of this feature. Mare Humorum appears in upper left corner of the photograph, made from an altitude of 2722 km. South is toward the top.

low interior after either an ejection of the mountain building material or a recession. If this be true, Eratosthenes, in its present condition, is a magnificent caldera. Its albedo is almost the same within the crater, at the rim, and on the mountain wall. The crater and its mountain disappear as completely at noon as do the typical mountain-walled plains. Plates 7–1 through 7–3 exhibit this characteristic difference between the two classes of ringed plains.

However, one must not be dogmatic concerning the hypothesis that the violence of an explosion caused the main differences between the two types of ringed plains. Craters which lack ray systems and excessive noonday brightening may merely be older. According to this hypothesis meteoritic or other dust has covered the rays and the craters. This result appears certain if there was sufficient difference in age. The author inclines strongly to the former hypothesis but he realizes that the truth may not be known until long after man has begun his explorations on the moon.

The positions of some of the largest, non-brightening ringed plains are marked on the accompanying photograph (Plate 12–2) of the full moon. Arzachel and Cyrillus are hybrids and possibly should have been omitted. Neison did not give the brightness of Aliacensis. The brightnesses of these craters compare with the general level of the bright mountainous areas, an average between 4 and 5, whereas the first list averages more than 7.

There are other craters which appear at first glance to be rather bright merely because they are located on the dark floors of maria.

It is difficult to classify some of the ringed plains with respect to brightness. The rings themselves are not especially bright, but each is associated with an intensely bright craterlet. The string of rather large rings extending southward from Abulfeda is notable for this characteristic. Abulfeda (4½, peak 7) has a bright central peak whose relationship to the ring appears not to be accidental. The next in line, Almanon (5½, craterlet 8+), has rather similar characteristics with a small, bright craterlet on its outer wall. The third of the string is Abenezra (5, craterlet 6), which is lost at full moon except for the bright craterlet to the right of it. Apianus (craterlet 8 on the left wall) follows with a very

bright craterlet. These are four members of an arc of seven large ringed plains (Plate 3–8) which form the left boundary of a tremendous, slightly depressed area that is bounded on the right by one of the great lunar scarps. The craterlets indicate at least trigger action in formation of the craters but the unusual alignment seems to contradict such a hypothesis. The best guess is that the rings do lie along a complex, perhaps Tychonian, fault which provided preferential conditions for the action of large meteorites. If this be true the seven are, for the main part, endocraters.

Mare Orientale, a vast circular formation on the lunar farside, is pictured in Plate 12–3 and appears on the Farside Map in Chapter 19. Underlain by a lunar mascon, or mass concentration, indicated by its effects on the orbits of lunar satellites, Orientale has several rings around its dark, mare-like center. Like the ripples that move outward from the spot at which a stone strikes the water, the shock waves of the impact that triggered Orientale formed the rings around it. Then lava flowed to fill the center and even some low terrain between the shock rings, as the whole formation developed.

The available evidence concerning non-brightening lunar craters does not distinguish with certainty in individual cases between impact and endo-lunar origin. Those features of this kind that are due to impact would seem to have been caused by meteorites or asteroids striking the lunar surface at far lower velocities than was the case in the great explosions. Of course, we must remember the possibility of dimming with age. The distribution of these craters favors an endo-lunar origin for most of them. Except for those marked by bright craterlets, almost none of these non-brightening, rather large craters with high external walls are found in mountainous regions, far from maria. Nearly all of them are either in or close to maria. In other words, they are found in the regions that provide the greatest evidence of flows of molten rock. Perhaps their relationship to the maria is somewhat similar to that of Copernicus and Tycho to their surrounding craterlets.

It would appear reasonable also that some of the craters that brighten only moderately under a high sun are endocraters for which the explosive forces were unusually large. Lines of such craters, as, for example, the moderately bright examples along the great ray to the left from Tycho to Mare Nectaris, appear to demand an origin other than direct impact.

The fact that such large volcanic craters are not found on the earth has no weight, either pro or con. We do not find craters like Copernicus here either. Surely great impact explosions must have occurred on the earth if they did on the moon, but the resulting very large craters would have been eroded by our atmosphere. It is possible, on the other hand, that no large explosive craters, either of the endocrater or the impact variety, may ever have existed on the earth. Crater-forming forces are comparatively localized and the moon is large enough that mere size does not preclude explosive forces there from equaling the terrestrial forces. On the earth an explosive force would have worked against six times the weight of similar masses on the moon, and tremendous craters may simply not have been produced.

The distribution of the largest of the violent explosion craters has led to an interesting and suggestive observation. It has been noted that these features tend to group themselves in the north preceding part of the moon. The diameter drawn on the accompanying Plate 12–1 shows that ten of the eleven greatest explosions belong to one hemisphere. Such a number is too small to do more than indicate a good possibility that the grouping is not accidental. But the chance of non-accidental grouping of these features is increased by the fact that during all the eons of time that the moon has kept one face toward the earth, the right limb is the one that has been the "leading edge" in its orbit around the earth. A very tentative guess toward an explanation, then, is that asteroids belonging to the earth-moon system, rather than to the sun, have played a very important part in the evolution of the lunar surface and have inevitably made their mark more conspicuously upon the exposed right limb, which has always faced them directly. The observed phenomenon of hemispheric concentration would have to be laid to chance only if these ringed plains were the results of impacts by solar asteroids. Despite the apparent logic of this explanation, the truth of the matter may eventually prove to be very different.

One other peculiarity that has been observed

is probably due entirely to chance. It may b worth mentioning, however, because data learne at a later time sometimes establish a cause an take the peculiarity from the field of chanc There is a great peninsula of bright, roug ground near the center of the visible disk. is conspicuous on any photograph of the fu moon (Plates 12–1 and 5–2). This peninsu breaks up what would otherwise be a tremendou mare covering most of the central part of ou side of the moon. The boundary between th two hemispheres which have been considere in the preceding paragraph is, insofar as can b determined, perpendicular to the axis of th peninsula, and must lie along the slope gradie between Mare Nubium at the right of the penin sula and Mare Tranquillitatis at the left. In gen eral it is approximately perpendicular to th "gashes" which are numerous in the area; se Chapter 15.

One more peculiarity between these pre ceding and following hemispheres of the moo must be recorded. The total mare area in th right half of the visible disk is several times tha in the left, and is particularly large on the lowe right side of the diameter discussed above.

The following conclusions concerning e plosion craters appear to be justified:

1. The largest craters, which brighten unde a high sun, were probably initially meteorit or asteroidal in origin, but assumed their prese form after subsequent endo-lunar action.

2. The small craters that brighten to starlik brilliance were formed by meteoritic impact.

3. Quite a number of non-brightening crate that exhibit bright craterlets are endocraters i itiated by meteoritic triggering shocks.

4. The large non-brightening explosion cra ters are probably in many cases endocraters of th caldera type, although it is at present impossibl to be certain in any specific case. The possibl dimming with age due to covering by meteorit dust must be remembered continually.

5. The craters found on the highlands o Sinus Iridum, on the Alps, the Caucasus, an the Apennines probably are mostly endocrater

6. There is a definite concentration of th principal explosion craters and of the maria i the portion of the moon that constitutes th leading limb as the satellite moves in its orb through space.

3 THE NATURE OF THE DOMES AND
SMALL CRATERS OF THE MOON

he small craters of the moon are classified by eison as craterlets, crater pits, and crater cones. ll of these categories will be referred to here s small craters or craterlets, although certain aracteristics could be used to divide them into bclasses. They vary in size from the smallest oservable object of this kind up to about five iles in diameter. The largest ones merge instinguishably with the larger craters. Domes e similar in appearance to some of the small aters, except that no more than a tiny pit is oservable in the top of these low, rounded hills at look something like human skin blisters (see late 7–2). The term endocrater will be applied, in earlier chapters, to any craterlet due to ny sort of internal pressure or explosion. This rminology is followed even when the fundaental but indirect cause may have been a earby, great impact explosion like the one that roduced Copernicus.

Any hypothesis concerning the formation of craterlets and domes must explain, or at least not contradict, the following data:

1. Many and perhaps all of the craterlets have external walls.

2. Lines of craterlets are common. Sometimes the lines are curved and form complex patterns. The craterlets in these lines do not brighten to a starlike appearance under a high sun.

3. Some of the lines are composed of both craterlets and domes.

4. Isolated craterlets are common on the mare floors. They generally appear to have floors deeper than the surrounding mare.

5. The typical isolated mare-floor craterlets have the same general appearance as the domes, except that the former have an opening in the top. These craterlets rather resemble collapsed domes.

6. Small craters are common around the rims of large craters and mountain-walled plains.

7. More than a thousand craterlets exist in

the Copernican area and at least hundreds near Tycho.

8. Some of the craterlets brighten very much under a high sun and become starlike. This is the same phenomenon observed in the cases of Tycho, Copernicus, and some others of the larger craterlike formations. Many craterlets brighten moderately. But still others tend to disappear entirely near noon.

9. Many of the craterlets that brighten have tiny ray systems. W. H. Pickering believed that all craterlets that become starlike have this cha¬ acteristic. Probably he was right.

10. Sometimes elementary rays from crate¬ lets combine to form the complex principal ra of such craters as Tycho and Copernicus.

11. In some cases lines of craterlets ha¬ coalesced to form valleys. Careful examinatio with large instruments reveals signs of craterle in many—perhaps in the majority—of the rills clefts. Rheita Valley is the result of such coale cence of larger craters.

Plate 13–1. Features of Deslandres outlined for use as a map. See Plate 8–5 for technical details.

12. Sometimes even fairly large craters form line. A beautiful example is the north-south line that touches the left hand wall of Pitatus (see Plates 13-1 and 13-2).

13. An arc of craterlets may enclose a partially sunken area. Excellent examples can be found (Plates 13-1 and 13-2) south of Hell and in the lower left hand portion of Lexell.

14. Many small craters can be observed on the Apennines (Plate 12-3) under favorable conditions. Somewhat larger ones are observable on the highlands of Sinus Iridum. The Alps and the Caucasus also show such craterlets.

Craterlets will be discussed here under the following four classifications:

1. Craterlets near large explosion craters; 2. craterlets on or near rims and floors of craterlike formations; 3. lines of craterlets and domes; 4. craterlets that become starlike under a high sun.

1. Plate 7-1 shows the region left of the great explosion crater Copernicus under a very low sun. Eratosthenes is at the lower left of the picture. The ghost ring, Stadius, is below the center. There are hundreds of craterlets forming all sorts of patterns. Even Stadius is, in part, a circle of craterlets. In some places craterlets coalesce to form rills, or narrow valleys. Some of these valleys are radial to Copernicus, but others follow other directions in various ways.

Two ovals of craterlets touch Stadius at a common point on the south, with the smaller oval inside the larger. Under the very low morning sun of this picture the Copernican rays are faint. However, an examination of Plate 7-5 reveals that two oval rays shown there coincide with these oval loci of craterlets. Clearly there is a close relationship between craterlets and the components of major rays—at least on the basis of this evidence.

A little to the left of Copernicus and considerably south of the center of Plate 7-1 is a craterlet of the type common on the floors of the maria. To the left of it is a dome. Just to the south of this crater and dome and extending on the right to an area directly south of Copernicus is one of the three large, rough, dark regions associated with Copernicus and almost unique on the lunar surface. The picture shows about a dozen domes in it. A number of craterlets and small domes can be seen among the hills of the far southern area of the picture. Such domes are too low to be observed when the sun is high above the horizon.

Plate 7-5, extending to the right of Copernicus, does not show as great a mass of craterlets as are visible on the other side, except to the lower right, where they tend to form radial valleys. A beautiful group of five domes is just to the lower left of Hortensius in Plate 7-2. An excellent single dome is shown right of Milichius, and two large, low domes are seen a little farther north of it. There are two tiny pits in the larger of the pair. Several other of these larger domes as well as numerous smaller ones can be seen on the plate.

2. Plate 8-6 was made to exhibit details of Tycho and of the floors of the nearby walled plains. The many craterlets on the flattish floors of mountain-walled plains to the lower left of Tycho tend to form rills radial to that crater. These craterlets are for the most part smaller than those left of Copernicus and there are fewer non-radial rills. On the lower left part of the floor of Tycho is a beautifully curved chain of craterlets approximately parallel to the edge of the floor. To the lower right of Tycho can be seen a large, somewhat sunken oval region composed of small craters, most of which are larger than craterlets. This area breaks the rim of Tycho, and has the distinct appearance of a sinking area that stopped subsiding too soon to be listed as a walled plain. It certainly is newer than Tycho. Rim craterlets may be noted on Tycho as on all large craters.

3. Plate 10-2 shows among other things the Straight Wall, the old walled plain within which it lies, a rill which formed along a line of craterlets right of the wall, and a line of craterlets and domes perpendicular to the wall and reaching it from the left, better seen on Plate 10-3. Such mixed lines of craterlets and domes are important in an interpretation of their nature. The irregular spots in the wall and the craterlets at its ends should be noted.

Important features have been outlined on Plate 13-1 to make possible its use as a map for Plate 13-2. On the lower right hand floor of Lexell is a partly coalesced line of craterlets designated as A. The lines C and D border a valley that extends from the interior of Lexell to

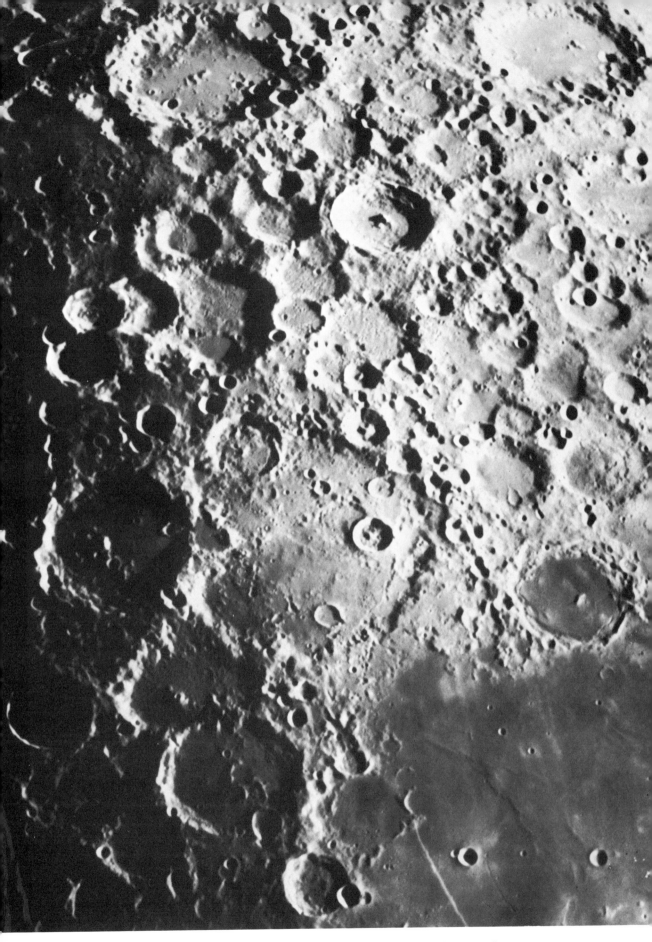

Plate 13–2. The unusual detail in and near Deslandres (see Plate 13–1).
See Plate 8–5 for technical details.

he lower right to "Cassini's Bright Spot." *H* is a long arc composed of both domes and craterlets. Several groups of domes can be observed north of it with a few craterlets mixed in. The line *I* is an arc of craterlets partially surrounding an area that has sunk below the general surface. *O* is a line of craters somewhat larger than craterlets. The crater lines *O-P-Q* bound a triangular area just to upper left of Pitatus. Numerous other lines of craters exist in the area.

4. Any good photograph taken near full moon (Plate 5–2) will exhibit some of the random craterlets that brighten to starlike appearance under a high sun. They appear to be the results of explosions too violent to have originated on the same rather quiet moon that produced the maria and mountain-walled plains, and it seems best to attribute their origin to meteoritic impact. Others, which brighten to a lesser degree, may result in some cases from impact but in other cases may be the product of unusually violent endo-explosions long ago.

The following conclusions concerning craterlets and domes appear to be warranted:

1. Volcanic craterlets near the great explosion craters appear, for the most part, to be a result of those explosions, even though the original explosions that formed the large craters may have been due to impact. Shoemaker and others consider such craters to be due to impact.

2. In many cases such craterlets must have formed along faults or along and within rills.

3. The craterlets near the rims of large craters and mountain-walled plains are volcanic and indicate faults in the surface rock.

4. Mixed lines of craterlets and domes exist, indicating a common generic nature.

5. Possibly the domes are bubbles formed by internal gas (conceivably including steam) pressure. Or a tube of molten rock may have elevated the surface locally, especially in connection with phase changes of minerals in the lunar mantle.

6. Most of the isolated small craters on the mare floors are endocraters.

7. Probably the typical rounded, isolated craterlet of the mare floors is the result of the collapse or bursting of domes. The domes look like the mounds that form on the surface of a gently boiling pot of cornmeal mush (Plate 13–3). Many isolated mare craterlets look like the remains of such collapsed "mush domes."

8. The origin of the isolated craters that become starlike under a high sun is probably meteoritic impact.

9. Some of the isolated craterlets that merely brighten moderately are probably due to low-velocity meteoritic impacts. Others of this kind are very probably endocraters. Not enough data are available to establish definite criteria for individual craterlets of this type.

10. Many rills, or clefts, show endocraterlets.

Plate 13–3. Volcanic domes, cones, and ridges, photographed by Lunar Orbiter V from a height of 112 km near crater Marius. Two sinuous rills cut across the main ridge; the head of one is in a nearly circular depression, the other originates in a long spearhead. Two kinds of domes shown are low, smooth, gently rising domes, and steep-sided, rugged, heavily cratered domes.

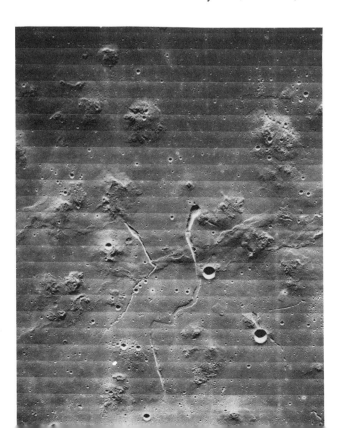

14 THE NATURE OF THE LUNAR RAYS

Much of the interpretation of lunar features is controversial, but nothing is more so than the rays and the ray systems. Under a low sun they become inconspicuous, but at full moon they demand attention from even the most casual observer.

The following data are generally accepted by students of the rays:

1. There is very strong tendency for ray systems to be conspicuous only under a high sun.

2. The systems of bright rays are related to crater-like features that brighten under a high sun in a manner similar to that of the rays themselves.

3. In major systems short, "plume"-shaped rays are common, usually with their axes radial to the parent crater.

4. The long, major rays are complex and usually are not exactly radial.

5. Conspicuous ray systems are not necessarily related to large craters.

6. Many short, plume-like single rays issue from bright craterlets or at least from bright spots.

7. Often, perhaps always, the rays appear to be some sort of surface marking, although they may lie along scarps and ridges.

8. A ray appears at close to the same co longitude each month.

9. Rays are observable in all types of topog raphy. However, a bright, rough surface make their observation more difficult for observers or earth.

The hypotheses concerning the nature o the rays fall into two general groups:

1. The rays consist of, or at least resul directly from, ejecta expelled by the paren craters.

2. The rays are due to gaseous material or t dust that issued from craterlets or from fault in the ray area. These craterlets and faults ar related to the primary crater.

At present some form of ejecta hypothesis i favored by the majority of lunar students wh are trained in physical theory.

Five diagrams have been included in thi chapter as aids in identifying features of th rays. The 108 features numbered on the diagram were traced from the photographs to which eacl diagram is keyed. The numbers in parentheses following the names of these features in th general discussion, refer to the list. The alpha betical list of the features follows:

Table 5

1	Abenezra	55	Landsberg
2	Abulfeda	56	Lindenau
3	Agrippa	57	Marius
4	Albategnius	58	Maurolycus
5	Almanon	59	Menelaus
6	Alphonsus	60	Milichius
7	Archimedes	61	Mösting
8	Aristarchus	62	Oval, large
9	Azophi	63	Oval, small
10	Bessarion	64	Parry
11	Bessel	65	Piccolomini
12	Biot	66	Pitatus
13	Bonpland	67	Pitiscus
14	Brayley	68	Plato
15	Bullialdus	69	Polybius
16	Byrgius	70	Polybius A
17	"Cassini's Bright Spot"	71	Polybius B
18	Catharina	72	Polybius K
19	Clairaut	73	Pons B
20	Clavius	74	Pons C
21	Copernicus	75	Pons E
22	Copernicus	76	Posidonius
	bright spot	77	Proclus
23	Cyrillus	78	Ptolemaeus
24	Darney	79	Pytheas
25	Deslandres	80	Rectangle,
26	Eratosthenes		unnamed
27	Euclides	81	Reichenbach
28	Euler	82	Reinhold
29	Fabricius	83	Rheita
30	Flammarion	84	Rothmann
31	Flamsteed	85	Seleucus
32	Fracastorius	86	Sinus Aestuum
33	Fracastorius E	87	Sinus Iridum
34	Fra Mauro	88	Snellius
35	Furnerius	89	Stadius
36	Furnerius A	90	Stevinus
37	Gambart A	91	Stevinus A
38	Gassendi	92	Stiborius
39	Gay-Lussac	93	Stöfler
40	Geber	94	Tacitus
41	Gemma Frisius	95	Theophilus
42	Godin	96	Timocharis
43	Goodacre	97	Tobias Mayer
44	Guericke	98	Tobias Mayer C
45	Herodotus	99	Triesnecker
46	Herschel	100	Tycho
47	Hind C	101	Vlacq
48	Hortensius	102	Wichmann
49	Janssen	103	Zagut
50	Kaiser C	104	Zagut B
51	Kepler	105	Zagut D
52	Kepler A	106	Zagut L
53	Lalande	107	Zagut P
54	Lambert	108	Zagut R

Plates 14–1 and 14–2 show that a major ray system originates at craterlets just outside and to the right of the hybrid mountain-walled plains Furnerius (35) and Stevinus (90). The first unusual feature of the system is the trapezoid of rays shown quite well on Plate 14–2. On Plates 4–2 and 4–3 the low morning sun reveals the craterlike features but the rays have not yet become visible. On Plate 4–4, at phase 5.35 days, the sun is low enough so the rays do not disguise the general topography. The Stevinus craterlet lies on the longest side of the trapezoid, which begins at Snellius (88) and continues southeastward, touching Rheita (83) and Metius at their left rims. This extended side includes a small, bright crater, in Janssen (49), which has a distinct ray system of its own. At this point it meets the end of one of the great western rays from Tycho (100). Farther south the ray becomes much less distinct but under favorable lighting can be traced to a bright craterlet inside the southern wall of the ringed plain Vlacq (101). Here it appears to end, as does also the least intense of the Tychonian rays to the left.

Another principal ray, apparently from the Furnerius craterlet, ends at Biot (12). The last major ray of the system is long but weak in intensity. It belongs to both the craterlets. It first cuts the northern rim of Reichenbach (81), then enters Mare Nectaris and cuts the northern rim of Fracastorius (32). To the right of Nectaris it cuts the area common to the rims of Theophilus (95) and Cyrillus (23). If the ray is composed of ejecta, these tangencies would appear to be strange. All the rays of the system are conspicuous loci for small craters.

The southernmost of the great rays to the left from Tycho is weak and ordinarily escapes notice, but it can be traced faintly on Plates 5–2 and 14–2, ending indefinitely, with the ray from Stevinus, at the bright craterlet in Vlacq (101). The path of the next of these rays counterclockwise, and its ending on the Stevinus ray at the bright crater in Janssen (49), are definite. Certainly this pattern is not to be expected through chance. The third of these rays is directed toward Stevinus itself. Its origin is at the large mountainwalled plain, Stöfler (93), and it cuts the northern wall of Maurolycus (58).

The most intense of the rays to the left from Tycho exhibits so much evidence concerning the nature of rays, because of both its position and

Plate 14–1. For use, aided by a diagram, in identifying features related to the rays. See Plate 4–8 for technical details.

its structure, that it is worth describing in detail. Plate 14–1 is chosen to show its position at a low enough sun to permit easy identification of craterlike features. Plate 14–2 provides a normal

Figure 12. Map for Plate 14–1.

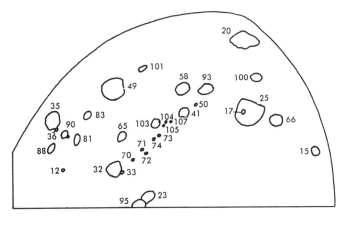

view of it as seen just before full moon. Plate 8–4 a day after full moon, was made on an extremely contrasty plate to emphasize certain details of its structure.

For a large part of its length the ray is double the components separating approximately at the northern rim of Stöfler (93). The southern branch passes just south of the starlike crater Kaiser C (50). That seven-mile-diameter crater has a nice ray system lying within this major ray. Next the ray cuts the southern rim of Gemma Frisius (41) It continues to the left through a bright craterlet and the 18-mile-diameter Zagut B (104) with its smaller flanking craters Zagut P (107) and Zagut D (105). Beyond this it meets bright Zagut R (108) and a very bright unlisted crater. Both of these lie on the northern rim of Zagut (103) Right of Zagut the ray is marked by two bright, small craters, the first of which lies on a pronounced scarp that runs from the Altai Scarp to the lower right rim of Zagut and beyond. Examination of this southern branch on Plate 8–4

ows that it is actually a linear series of bright
tches.

The northern branch first attracts attention
the vicinity of Kaiser C (50), which lies be-
een the two branches. It cuts the northern
n of Gemma Frisius (41) but remains much
nter than the southern one, even suffering
long interruption, until it reaches the bright
aters Pons B (73) and Pons C (74). Left of
ese it is almost non-existent until it reaches
ns E (75), 16 miles in diameter, just right of
e Altai Scarp. To the left of that scarp it is
arked by two craters on a short north-south
e: Polybius K (72), which has an extremely
ight craterlet on its lower left wall, and Polyb-
; B (71). Following to the left is Polybius
(70), which is ten miles in diameter and
ssesses a small ray system that lies within
e major ray. The last crater to mark the great
is Fracastorius E (33), nine miles in diameter,
the lower right rim of Fracastorius (32). The
y ends just beyond this at a fainter crater

that marks the junction with the long ray from
Stevinus. The Tychonian ray, the one from Stevinus, and a scarp prolonging the outer upper left
hand wall of Fracastorius to Piccolomini (65)
form a conspicuous "sling" that appears to sus-
pend Fracastorius. The occurrence of these combinations through mere chance, which would
necessarily follow from any ejecta hypothesis,
would be extremely improbable. Most of the
brightness of the great ray results directly from
the features that have been described along its
path. For an ejecta hypothesis to carry convic-
tion, it should present a reasonable explanation
for the initiation of all these craters through
ejecta from Tycho. In the area south of the great
ray and right of the Stevinus system there is only
one bright crater that does not lie on a Tychonian
ray. It is Neander N, diameter eight miles, with
a beautiful little ray system.

The relationships listed above appear to be
reasonable under the following hypothesis: The
major explosion that created Tycho shattered

Plate 14–2. Interrelationships of Tycho-Furnerius ray system. See Plate 5–2 for technical details.

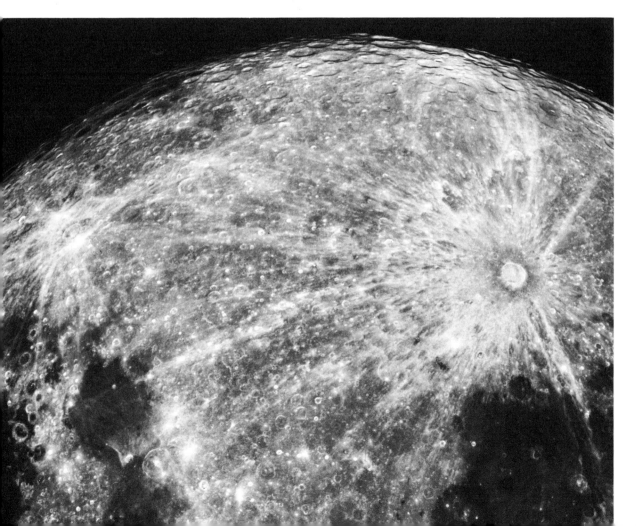

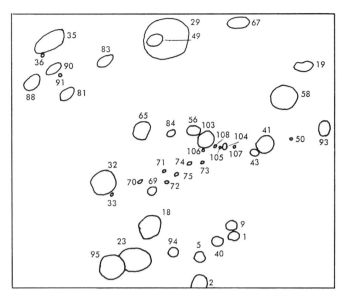
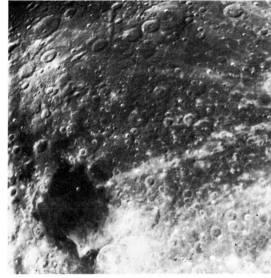

Figure 13. Plate 8–4 and map for Plate 8–4. An enlargement of this photograph is in Chapter 8.

the surface rock for a long distance. This produced the nimbus area which, in the case of Tycho, is dark in the inner part and bright in the outer. We see this type of effect quite commonly around the hole drilled by a bullet through a pane of glass. Near the hole there is a nimbus with too much shattering for any pattern to be observable. Beyond this nimbus there is a strong tendency for radial cracks. At the time of the Tycho explosion, many hundreds of small craters formed in and near the nimbus. Gases, escaping from the parent crater, constituted a very low-density outward radial wind and carried the ejected gases and dust from small craters as radial plumes away from them. Such plumes are observed by the hundred, especially in the Copernican area where the original surface was much smoother and darker than that around Tycho.

Wherever there had been previous formation of strong explosion craters, each would have produced, on a smaller scale, the same type of effect as did Tycho, with radial faults extending beyond their small nimbi. Whenever such a minor system had been formed in the vicinity of a still earlier one, the principle of least action would produce a tendency for those of its faults that happened to lie in the direction of the older crater to merge with the latter's faults and reinforce

them. Perhaps the finest example of such operation is the Stevinus-Furnerius major system. Here two very intense craterlets, re forced by the perimeter faults of Stevinus a. Furnerius and receiving considerable assistan from those of Snellius, Rheita, Fracastorius, Th ophilus, Cyrillus, and other pre-existing crate teamed to form a major system during the p Tychonian era. The same type of reinforceme found between neighboring explosion crate would take place when a violent explosion curred near the extension of a side of a mou tain-walled plain. The newly formed syste would tend to reinforce the fault containi the side of the older feature. One of the ni examples of such reinforcement is found in t tangencies of the two branches of the inter Tychonian ray with Gemma Frisius (Plates 14 and 14–2). A perfect example of fault reinforc ment by neighboring craters is the southe branch of this ray. Another is the line in t northern branch beginning with Pons B and en ing with Fracastorius E. It appears very difficu to ascribe these five craters directly to ejec from Tycho; on the other hand it seems plausib that when the overshadowing Tycho explosi occurred, the lesser, pre-existing systems me tioned here modified its effects to produce t effects described above.

Evidence has been presented that pre-existing
ults, especially those that are roughly radial
a major explosion, tend to be opened more
dely and therefore are apt to be loci for crater-
s. In such radial cases the outward plumes
n reinforce each other and bring about the
mation of a complex major ray. The intensity
the complex ray would be a function of the
ount of discharge from the craterlets. Plate 10–
hows a beautiful example of such a result. A
nounced scarp runs southward from the shore
Mare Nubium starting at a point right of
atus. For part of the path near the mare, the
rp is double. It is directed toward the right
nd portion of Tycho, near which it divides
ain to continue as a "V" almost to the rim of
avius. Under proper low lighting it is fairly
sy to trace the whole path. Plate 10–1 was
posed at a phase long enough before sunset
that both the scarp and Tycho's double ray
e observable and the rather close coincidence
n be noted. The scarp certainly cannot have
en formed by ejecta from Tycho. The right
nd ray in this region contains an excess of
aters of various sizes, which are visible along
path in the mare. The mare section of the
nter left component, however, deserves the

more study. Until it reaches the craters just south
of Bullialdus, it consists almost entirely of five
brightish streaks. Near the center of the strongest
of them is a conspicuous double craterlet. Craterlets exist in two of the others and are easy to
visualize in the remaining two. To be sure, the
craters could be explained by either of the two
ray-formation hypotheses, for one might, with
Shoemaker, think of each craterlet as having
been formed by an ejectus from Tycho. The
ejectum hypothesis, however, must accept pure
chance as the reason for the correspondence of
the ray with the scarp.

When we examine the Copernicus-Kepler-
Aristarchus triangle (Plate 7–6), we find the
same strange coincidences which were observed
on a larger scale in the Tycho and Stevinus-
Furnerius systems. One complication partially
masks the relationships: Major rays cannot be
observed within the shattered nimbi of great
explosive craters. In the Tycho-Stevinus oversystem there was a great distance through which
we could trace the paths of four connecting
rays. In this second example the distance between the nimbi of Copernicus and Kepler is
only a third of that between the craters themselves. Although a slight study is possible in

Figure 14. Plate 7–6 and map for Plate 7–6. An enlargement of this photograph is in Chapter 7.

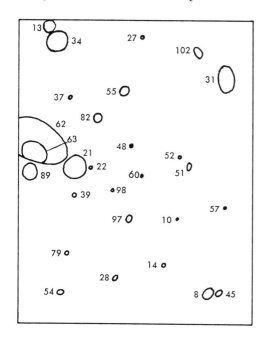

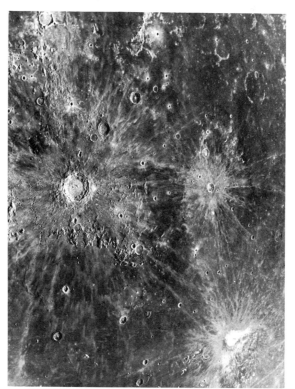

this region, it is only the rays that also involve Aristarchus (8) that give definite evidence concerning interrelationships. The brightest spot (22) on Copernicus is found on its outer right wall. The spot is not listed in the IAU catalogue and is not mentioned by Neison, Goodacre, or Wilkins. At colongitude 38° it appears as a single craterlet, but it is actually complex. To the right, as soon as the badly shattered area has been passed, we observe a complex ray leading directly away from it toward Kepler A (52).

Three major rays lead in the general direction from Copernicus toward Aristarchus. One of them is observed first at Tobias Mayer C (98) or perhaps at the very bright nearby double craterlet Tobias Mayer H. Next it passes through a bright craterlet (perhaps K but not listed in the IAU catalogue) with a small ray system. The ray finally merges with a ray from Aristarchus in the area north of Bessarion (10). The next ray (north) connects Tobias Mayer (97) and Aristarchus. The last of three appears first to the north of Tobias Mayer and extends to Brayley (14) where a ray from Aristarchus takes over.

The rays between Kepler (51) and Aristarchus are the best examples. The two craters are connected directly by two narrow rays, each which extends the whole distance. Just to t right of this pair a ray from Kepler ends Marius (57), as does a ray from Aristarch When one observes these last rays and lets eye continue the arc through Euclides (27) Bullialdus (15) and Tycho, Plate 5–2, he mu steel himself to accept the arrangement as m coincidence! To the right of Aristarchus the sa type of coincidences connect these craters Olbers and Seleucus (85) (Plate 11–4). It shou be observed that the strong ray from Seleuc forms the northern boundary of the much d turbed area north of Aristarchus. This becom natural if the ray follows a great fault. It wou be similar to the situation that causes the gr ray from Tycho to end at the right hand r from Stevinus.

Even without the data from the Tych Stevinus system, these coincidences suggest th the positions of rays from one crater sometim affect those from another. However, by the selves the rays of the Copernicus-Kepler-Arist chus triangle cannot be considered sufficie proof. The coincidences may be due to chan

When we consider the most elementary c tail of the ray systems of Copernicus, Plate 7-

Figure 15. Plate 5–2 and map for Plate 5–2. An enlargement of this photograph is in Chapter 5.

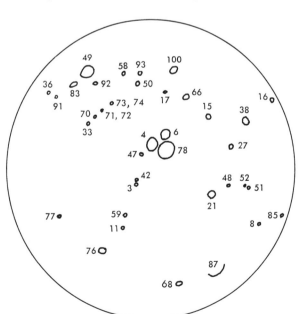

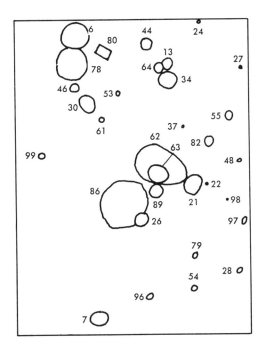
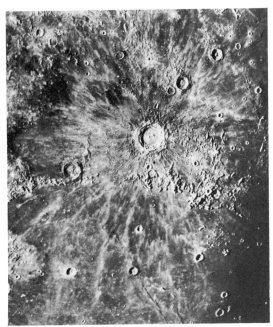

Figure 16. Plate 7–5 and map for Plate 7–5. An enlargement of this photograph is in Chapter 7. The line drawing cannot show variations of contrast that are in the photograph. For example, number 21 is the most conspicuous in the photograph, and 86, although important, barely shows in the photograph.

and Tycho we are impressed by the profusion of the plumes mentioned earlier. Copernicus exhibits more of them than Tycho but that may be because the different pre-existing surface affects observations. We observe the plumes as pointed toward the great crater. Usually the point is somewhat brighter than the plume. Sometimes a craterlet can be seen; it can be assumed that a craterlet is always present. The major rays appear to be composed of these plume elements and at some places in them the plumes are easy to disentangle. In the most intense ray to the left from Tycho, the plumes often arise from fairly large craters. Any pre-existing craters, insofar as this hypothesis of ray formation is concerned, may be either endocraters or impact ones. According to the non-ejectus hypothesis, craters formed at the time of the parent explosion are endocraters resulting from the general disturbance brought about by that explosion. Probably these craters lie along faults and in at least many cases are similar to blowhole ones. Such faults will be of two kinds; faults that antedated the great explosion and have merely been strengthened by its shock waves in the surface rock, and new faults caused by the shattering waves which were strongest in the nimbus and decreased radially. The direction of the plumes would result from a low-density stream of gas issuing like a radial wind from the site of the parent explosion. The cause of the initial explosion is immaterial. The only requirement is that it must have been violent enough to have modified the surrounding surface.

In the cases of the most violent explosions, such as Tycho and Copernicus, the rock was shattered for a considerable distance and faults were opened for hundreds of miles farther. The hypothesis of such an origin demands that lines of craterlets be more common than would be expected through chance. Wherever the direction of the fault approximated to the radial, the plumes necessarily pointed somewhat along the fault line and reinforced each other. The result was a composite great ray in those cases where there were sufficient craters along the fault. The two most intense of the northward rays from Copernicus are among the best examples. In the case of the left hand member of the pair (Plates 7–2, 7–5, and 7–6) we observe the famous line of small craters extending northward from Stadius (89) until it merges with the beginning of that ray. These craters, especially at the northern end of the line, have overlapped sufficiently to form rills, and the rims cast observable shadows when the lighting is right. Within the ray itself we observe a ridge that finally turns right to merge with the second great ray, and continues within it to the ringed plain Lambert (54). Unlike the first great ray, the second one is almost radial. Its southern extension lies in a much rougher area than that of the first ray and is not apparent, but once the background farther north permits the appearance of the ray to become definite, the line of craters all the way to Lambert is conspicuous. The area in Mare Imbrium to the right of the second ray is worth many hours of visual observation and photographic study. Throughout this area the head craterlets of the plumes line up in a beautifully complex pattern. Probably the pattern shows so plainly because it is imposed on the comparatively smooth surface of the dark mare. We can expect manned exploration of the site to show that most of the faults are cracks which resulted from the original cooling of the mare floor and which were opened wider by the shock wave of the Copernican explosion. Lest there be some misunderstanding, it must be stated that there were almost certainly some ejecta from the formations of the greatest craters. The question under discussion is one of dominance. It probably is true that nimbi, such as those of Tycho and Copernicus, owe their present appearance principally to ejecta.

To the left and upper left of Copernicus, beyond the shattered nimbus (Plate 7–5), is a smaller but even more striking area of ray pattern than the one in the mare. Within this area is Stadius (89) which today, whatever it may have been in the past, is little more than a ring of small craters. If previously the formation of the mare had reduced it almost to a "ghost" condition, the Copernican explosion would have tended to produce such craters along the faults of its former rim. Touching Stadius on the south is the mutual point of tangency of two elliptically shaped rays (62 and 63). Their axes extend to the upper left away from Copernicus and the right hand end of the larger is lost in that great ringed plain. Craters are conspicuous around the perimeters of the ovals; it is possible to count at least 57 small craters around the larger one. These ovals are merely the most conspicuous features of a complex pattern. Many individual plumes can be seen in the area. Straight complex rays, quite commonly originating on one or the other of the ovals, can be seen in about a dozen positions. The pattern certainly is non-accidental and just as certainly it originated with the shock waves of the Copernican explosion. It would appear that pre-existing faults were opened wider and that craterlets formed along them. At first study the oval forms are unexpected. But the area of the larger corresponds well to that of Sinus Aestuum (86) and the shapes also match quite well. The end of Sinus Aestuum toward Copernicus has become ghostlike. It is bounded by Stadius and an arc of the larger oval (62). Quite possibly the two areas are old formations of a similar type. If so, the only difference is that in the case of the oval ray, the formation had become entirely a "ghost" before the birth of Copernicus, which merely opened more widely the faults around the old rim.

The non-ejectus hypothesis accounts nicely for the observed features. We must, however, consider the recent work by Shoemaker. This geologist has made a wide study of terrestrial craters, including Meteorite Crater and manmade explosive ones. He has studied the ejecta from Meteorite Crater and has identified them with respect to positions of origin. In addition to observing these data he has made calculations of the effects of the dynamic forces involved. Following these terrestrial observations he has made measurements of the positions of features connected with the Copernican ray system, and through his calculations has come to the conclusion that they can be accounted for by the ejecta hypothesis. This includes even such features as the lines of small craters and the oval rays. His results appear to be reasonable enough, and if the Copernican data were the only ones available it would be impossible to discriminate between the hypotheses. We would be forced to wait for on-the-site observations.

Plate 10–4 corrects foreshortening at the center of the plate. Its most valuable contribution lies in the interpretation of the features to the right of Tycho for which normally there is much foreshortening and also a large variation

amount. The most important feature shown
the great right hand gap of roughly 130° in
ray "pinwheel." Proclus, in another part of
lunar surface, exhibits a very similar peculi-
ty. Within this gap is found Mare Humorum.
the south and left of Humorum are found
merous marelike areas, some of them huge.
en the rough parts of the area are less bright
an are such regions in general. This region usu-
y is thought of as an extension of Mare Hu-
orum but it would probably be better to con-
er both it and the mare as extensions of
eanus Procellarum. There is little question but
at these, with Mare Nubium and at least parts
Mare Imbrium, are generically one tremendous
ean. The alignments of craters do give hints of
o rays within the region south of Humorum
hough there is almost no evidence of ray
ightening between the craters. One of these
ts the southern shore of Mare Humorum and
ntinues toward the very bright explosion crater,
rgius, in the same way that one of the rays
the left from Tycho has been described as
inting directly toward Stevinus. Byrgius has
considerable ray system of its own. The cause
the two pinwheel gaps is independent of
ether one considers an ejecta hypothesis of
y formation or one due to faulting. In each
pothesis the gap must be due to an extremely
gh temperature in the area either at the time
hen the ray system was formed or later. Per-
ps Mare Humorum is younger than Tycho.
rhaps it is older but this whole area had not
t had time to cool sufficiently when the great
plosion took place. Every serious student of
e nature of the lunar surface should make a cor-
lated study of this area, the one to the right
Proclus, the similar one above and to the left
Mare Crisium, Mare Australis, and the region
the left end of Mare Frigoris.

One other Tychonian ray (Plate 5–2) must be
scussed, not because of its intrinsic impor-
nce but because of the numerous incorrect
atements that have been made concerning it.
has been considered as extending to the north-
n limb and even has been spoken of as the
ound the moon" ray. It is not observable in
e area near Tycho. The first evidence of it
at "Cassini's Bright Spot" within Deslandres.
he ray passes the left hand rim of Albategnius
4) and continues northward to the small ringed
lain Hind and the smaller but brighter Hind C

(47). North of Hind the ray is not seen again
until the shore of Mare Serenitatis is reached, although the illusion of a ray is preserved by the
line of ringed plains Godin (42), Agrippa (3),
and Menelaus (59) which is the brightest object
along the locus. From Menelaus a ray proceeds
northward in Mare Serenitatis to Bessel (11).
Although Menelaus does have a small ray system, this ray may belong primarily to Bessel.
It is the only conspicuous ray from Bessel and
its direction probably has been due to Menelaus.
Quite possibly the shock waves from Tycho
widened the fault between the craters. The
pseudo-ray from Tycho ends in Mare Serenitatis
but the pattern is taken up again by four bright
craters farther north. Much of this description
parallels that of the ray from the left of Tycho to
Fracastorius. That one also suffered gaps between
craters. However, those were shorter than the
long, complete breaks in this case. Of course,
it is possible that the existence of Godin, Agrippa,
and Menelaus is due to the shock waves from
Tycho. In such a case the absence of plumes
would be a result of their great distances from
Tycho: The outflowing gas stream from Tycho
could have had too small a density to affect
matters noticeably at that distance.

Plate 11–4, made from one of the Wright
globes, shows that the pattern exhibited by the
ray triangle of Copernicus, Kepler, and Aristarchus continues to the right. At the time the photograph used for this globe was exposed, librations
had carried the bright ringed plain Olbers too
close to the limb of the moon for examination,
yet its position is shown clearly by the rays that
extend to the left from it. Both Neison and
Goodacre consider it to have a ray system equal
to that of Kepler. The long ray to the right from
Kepler is directed toward it and the same is
true of the long ray extending in both directions
from the southern rim of Seleucus (85).

Any conclusions concerning the nature of the
lunar rays probably will remain controversial for
several years. The final truth may not be known
until after years of thorough, on-the-site exploration. Even today, geologists are much divided
concerning some of the terrestrial problems which
they have studied at close hand for a good many
years. As of now it would appear that although
ejecta from the large parent craters did play a
minor role in ray formation, small craters in the
ray areas were much more important.

15 PECULIAR FEATURES
OF THE SURFACE

One of the basic characteristics of the human mind is the deep urge—really a compulsion—to correlate. The search for correlations and interrelationships is one basis of scientific method and practice: We think of Galileo, Newton, Einstein, and Freud as being very great men because they contributed so much to our known systems of broad correlations. The same unifying tendency is certainly related to the search for a single primary cause of all things, and thereby to theology and metaphysics.

All of this holds true for little problems as well, such as the study of the surface of the moon. The search for correlation causes most current investigators of lunar problems to concentrate on the maria, the mountain-walled plains, and the explosion craters; all of these combine into broad categories with many members each, and thus give rise to fertile generalizations. Very few of the apparently accidental features intrigue most lunar astronomers.

This practice of concentrating on easily correlated features is proper if it is not carried to an extreme. But the peculiar and isolated features deserve consideration too. Someday, on further study, they may well fit into important pattern. "The stone which the builders rejected, the same is become the head of the corner."

Descriptions of some of the more notable unrelated phenomena are set forth below.

1. *Solitary mountains.* One of the rarest lunar phenomena is the isolated, high mountain rising from comparatively level terrain. There are not many on earth either. Indeed, from their very nature we would expect such mountains to be scarce on either body. Their formation would have required an extremely localized but tremendous force to extrude semi-molten rock (or cinders) that cooled rapidly enough as it was pushed above the surface to keep from collapsing under its own weight. The smaller lunar attraction would make such events somewhat less improbable on the moon than here. There seem to be at least two very fine lunar examples of this unusual phenomenon: Pico and Piton, Plate 3-2 and 10-19, which rise from the lower left floor of Mare Imbrium. Quite a number of low, isolated mountains or hills may also be observed.

The published figures for the heights of mountains and depths of craters on the moon are notoriously inexact. Heights are usually computed by measuring the lengths of shadows. The altitude of the sun above the true horizon can be computed easily for each point of the moon, and the height follows from a simple trigonometric relationship. In application one difficulty is that the sun is not a point and the exact position of the end of the shadow is hard to determine. A second problem arises from the fact that if the shadow terminates in a bright area it appears to be longer than if it ends in a dark zone. Another source of error is the roughness of the ground: if there is high ground nearby the shadows are too short, and if there are depressions they are too long. Theoretically the observer determines only the relative height of the feature being studied above the surface at the end of the shadow. But even this becomes inexact if the shadow ends on a slope either toward or away from the feature.

As a result of these difficulties any hypothesis depending on relative heights, such as the ratio between heights of walls and diameters of craters, must be applied with extreme caution. This trouble is dramatized by the case of Pico (Plate 10–20). Pico is a steep, high mountain standing alone on what appears to be a great plain. It casts a long, sharp shadow the length of which one might think should have been accurately determined long ago. Neison quotes Mädler's determination as a height of 7,060 feet, and Schroeter's, "from three fairly accordant results," as 9,600 feet. The discrepancy is 34% of the smaller estimate! In their recent book Wilkins and Moore give the height as 8,000 feet. Goodacre in 1931 stated merely that it "rises according to various authorities from 7,000 to 9,000 feet above the plain."

Piton, which is just as good an example, is situated in the same area, roughly a hundred miles south and left of Pico; its height is stated as 3,000 feet.

Systematic work on the heights of lunar features has been carried on under the auspices of the United States Air Force. Thousands of photographs have been exposed under time-lapse photography and measured by modern methods. These produced better charts for Apollo landings on the moon. Since then radar and laser measurements of lunar altitudes from earth, as described in Chapter 17, have yielded ever more accurate lunar charts.

In order to understand the nature of these isolated mountains, one should examine the whole area carefully on the best photographs made soon after sunrise and just before sunset. Three hills and a ridge can be observed south and right of Pico on Plate 10–19. Close examination also shows the ghost of a ringed plain somewhat larger than Plato. Pico lies on this ring. In other words, it is a remainder of a large feature—not a solitary phenomenon. The same story of origin is told, somewhat less clearly, by the environs of Piton. Apparently there was no truly unusual development here.

It is true that the central part of Mare Imbrium was modified by asteroidal impact (Plate 10–22), but all of its section to the left must have resulted from the sinking which is characteristic of all the maria. As described earlier, this sinking produced a great many features: the great Apennine Scarp (Plate 10–18); the "marshy" peninsula extending to the right from it; the marelike floors of Plato and Archimedes, whose rims remained above the sea level; the Caucasian (Plate 3–14) and Alpine (Plate 3–28) scarps; the "Bay of Rainbows" (Plates 10–20 and 10–24) and the ghosts of the area to the left. Pico and Piton are the principal survivors of a former mass of great craters and mountains. If there had been impact here, all these reminders of past glory would have been destroyed.

2. Valleys. Despite a wide range in their morphological characteristics, numerous scattered features, varying all the way from very narrow rills to the mountain-walled plain named Schiller, may be grouped under this heading and may be divided into five general categories:

(a) Rills or clefts, for example, Schroeter's Valley.

(b) The Rheita "valley" and the approximation to a valley cutting through the southern part of Snellius.

(c) The Alpine Valley.

(d) The parallel "gashes" or valleys in the Ptolemaeus area.

(e) Schiller.

These divisions are considered in the following paragraphs.

(a) *Rills.* The description given by Neison

nearly a century ago is so excellent that it would be difficult to improve on it, and it is therefore presented verbatim.

"There is one class of formations which have not been placed under any of the three great classes described above, though possessing features rendering them in the highest degree interesting, but which, from unknown nature, cannot well be classified. These are the *rills* or *clefts*, long, narrow, deep ravines, canals or cracks, usually straight, often branched, sometimes curved, and not unusually intersecting one another; extending for considerable distances at times, generally traversing, without interruption, mound, ridge or crater pit in their path, though occasionally deflected by some object, or interrupted by others, when it recommences beyond and proceeds as before. One of the most difficultly visible, they are also one of the most inexplicable formations on the moon, and little information as to their origin can be derived from their situation, which is most diverse, at times lying on the open plains without anything to indicate beginning or end, often running through the midst of mountains, or extending from a crater to the open plain; at others they appear to form an intricate network around a formation, or are situated on the floor of a walled-plain or ring-plain."

Rills vary greatly in breadth and in general their depth is less than their width. In many cases, such as the great rill in Alphonsus, the bottom is illuminated under a high sun. The rill to the lower right of Aristarchus (Plate 7–6) is so broad that it is commonly referred to as "Schroeter's Valley." Others are so narrow that they can be detected only with powerful optical instruments. It is a reasonable speculation that the first men who try to make serious use of the moon will be plagued by thousands of rills too narrow for terrestrial observation.

The distribution of rills over the surface of the moon is by no means uniform, a fact that can hardly be accidental. Rills are not found in the central parts of the maria or in the central parts of the smoothish floors of great mountain-walled plains. Certainly such a finding must be a definite clue to their nature. They are seldom found in the roughest areas either, although they are comparatively common in the somewhat smooth regions separating great mountain masses. The absence of these features in mountainous regions

is possibly due, at least in part, to difficulty observation.

The following paragraphs contain brief d scriptions of some typical rills and systems rills, selected from the approximately 2,000 su objects that have been observed.

Triesnecker-Hyginus-Ariadaeus System (Pla 15–1). This is one of the two greatest rill sy tems, extending from Triesnecker, near the le hand shoreline of Sinus Medii, to Ariadaeus, the right hand shore of Mare Tranquillitat Much of it lies in a broad, flattish trough sou of the Haemus Mountains and north of the gre mass separating the Maria Tranquillitatis a Nubium. It is almost at the center of the visib disk.

The shortest but most complex part of th system is related to the ringed plain Triesneck Rills are in contact with Triesnecker on thr sides, apparently being prevented on the rig by Sinus Medii. (Such a distribution is seen in number of rill systems.) Study of the shadow of the Triesnecker rills shows that they are n all of the same depth. Some craterlets can observed along them, but not as many as in so other rill systems. Two branches extend nort ward almost to the nearby rill that runs to t lower right and to the upper left from oppos sides of the conspicuous little crater Hyginus. proper phases the Hyginus rill is observable ev with small telescopes (Plate 15–1–A). In t closeup of Plate 15–1–B, the craterlets along t rill show it to be fault produced. Domes can seen on the floor of crater Hyginus, 7 miles wi

The longest of this group of rills is the o that rises near Ariadaeus. It was discovered Schroeter in 1792 and is very easy to observe. eight places it appears to tunnel through elevat masses. These may actually be dams formed mo recently than the rill, or they may be featur that resisted the opening of a fault valley. will be an interesting task for explorers a f decades from now to determine the exact planation.

Hesiodus-Ramsden-Hippalus rills (Plates 2, 10–1, and 10–5). This triple system is m closely related to Ramsden on the upper rig part of the moon. The principal member is a ve long, narrow right-left valley that begins at t outer wall of Hesiodus (Plate 8–2) and runs the right for a couple of hundred miles alm directly toward Ramsden. Near the middle of its course it is interrupted for a long distance by a rather large north-south mountain mass. It reappears to the right of the mass and proceeds as though there had been no interruption. It finally ends at a tiny, bright spot (probably a small hill) left of Ramsden. The complicated rill system in the neighborhood of Ramsden (barely observable on Plate 10–5) reminds one somewhat of the rills near Triesnecker.

From the northern side of Ramsden two conspicuous parallel rills (Plate 10–5) run northward. It is possible that one of them is a continuation of another short rill extending northward into Ramsden. A third rill parallels the first two farther to the right. These three are perhaps the most interesting of their kind on the whole lunar surface. They are lost almost immediately in a mountainous section but reappear regularly wherever there is a depression. All three terminate

north of Hippalus, near the left shore of Mare Humorum. All three are easy to observe.

Other rills that should be examined by every student are those on the left hand floors of Alphonsus and Arzachel (Plate 16–1). The one in Alphonsus is quite remarkable. It has several intensely black spots with one or more craterlets in each. This rill is considered in detail in the next chapter in connection with residual outgassing, which may occur from Hyginus also.

The regions near Tycho and Copernicus show the closest possible relationship between craterlets and rills. The area left of Copernicus viewed at sunrise (Plate 7–1) shows hundreds of craterlets forming patterns that become rills by reason of mere contiguity. These appear in Plate 7–2 also.

A general rule is that a craterlet may be found wherever a rill makes a sharp turn or crosses another rill. There are a very few apparent exceptions to the rule, but even in these

Plate 15–1–A. The Triesnecker-Hyginus-Ariadaeus rill system. See Plate 3–1 for technical details.

Plate 15–1–B. Lunar Orbiter photograph of Hyginus Rill. The dark materials south and east of Hyginus may be from outgassing.

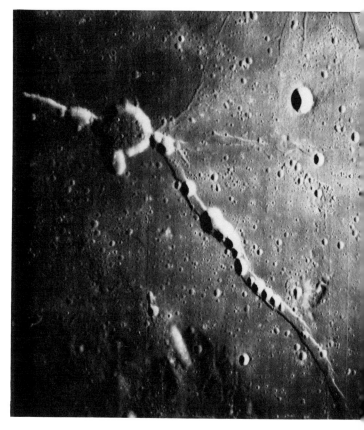

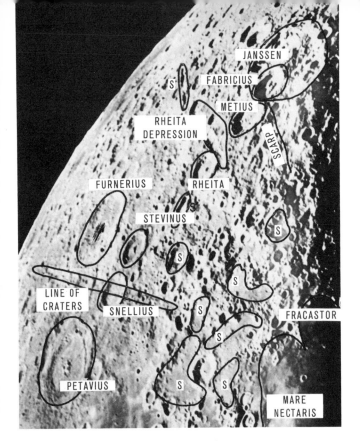

Plate 15–2. Features of upper left moon outlined for use as a map. See Plate 4–3 for technical details.

Plate 15–3. Details of upper left moon (see 15–2). See Plate 4–3 for technical details.

cases tiny craterlets may possibly exist. Moreover, there may well have been complete destruction of a craterlet during the evolution of broad rills.

There is also a rather noticeable relationship between rills and major rays. Both in general seem to follow faults, which usually exhibit what may be "blow-hole" craterlets or even, good-sized craters, Plate 8–4, as in the case of Tycho's great left land ray.

(b) *The Rheita Valley and its surroundings.* One of the two best known valleys of the moon, it is located in the upper left hand quadrant, which contains some of the finest examples of sunken areas to be found outside the maria. Plate 15–2 shows part of the region marked to identify the most interesting features. Plate 15–3 is from the same photograph. Some of the numerous unnamed sunken areas are designated by the letter S.

Plates 15–4 and 15–5, made respectively near sunrise and sunset, demonstrate that the "valley" is merely a deep trench along the left hand part of a depression so wide that it could be con-

sidered a shallow mountain-walled plain. Plate 15–4, made at phase 4.1 days with grazing sun light, reveals that the valley itself is compose of several confluent craters of varying sizes whos mutual walls have mostly collapsed. Except fo these craters, no sign of any sort of explosio either of internal or impact origin, is observab Directly north of Rheita on this plate the bare rising sun reveals a very large and rather squar depression with walls so low that it disappears less than a day; it is more difficult to observe tha even Deslandres.

Rheita Valley comes into fine focus in Pla 15–5, made by Lunar Orbiter. Details of the va ley, which have to be inferred in the other phot graphs, here leap into clear view. The broad va ley contains larger craters than those of Hygin shown in Plate 15–1–B, but Rheita appears have been formed principally by collapse in t photograph, with the craters taking part in, following the formation of, the valley.

Plates 15–2 and 15–3 show a long, narro right-left, slightly sunken area shaped like cigar, crossing the southern part of Snellius.

(c) *The Alpine Valley.* The origin of this great valley, connecting the Maria Frigoris and Imbrium, has been the subject of wide differences of opinion. It runs in a direction slightly up to the right toward Imbrium for about 75 miles, with a width of about four to six miles.

The top photograph of Plate 15–6 shows the valley under a morning sun, the second under an afternoon sun. The third (an early morning view) was made by infrared light at the 60-inch reflector. This feature begins as a narrow valley touching a dome and a crater. It then turns and, after somewhat less than a quarter of its length, suddenly widens. The Alpine Valley definitely appears to be a sunken area between faults, a graben, unlike the Rheita Valley and many of the rills. The Orbiter photograph in Plate 15–7 reveals that the Alpine Valley itself contains a sinuous rill for nearly its entire length. The photographs seem to contradict two of the suggestions that have been made in explanation of

its origin: that it is the result of a confluent line of craters and that it is a valley plowed out by some huge meteorite.

(d) *The parallel valleys in the Ptolemaeus area.* There are numerous narrow valleys in this area (Plate 11–1) that have been interpreted by some able students as gashes cut by fragments of an asteroid assumed to have struck first in the Mare Imbrium area. The principal one of these markings is shown in Plate 15–8. It angles up to the right of Ptolemaeus, roughly 150 miles long, and ends in the right wall of Alphonsus.

It is suggested that a very careful study be made of this plate which shows the structural details of the principal member of the system. It hardly seems possible that this can have been plowed out by any sort of missile.

The best guess today is that the "gashes" were opened along faults at the time of the sinkings that produced both the Maria Imbrium and Tranquillitatis.

15–4. Note unnamed, large, squarish depression terminator north of Rheita. Has general characistics of very shallow mountain-walled plain. Plate 4–2 for technical details.

Plate 15–5. Rheita Valley from Lunar Orbiter IV, from an altitude of 2972 km. A number of features in this photograph, in addition to Rheita Valley itself, appear to have collapsed.

Plate 15–6. (Top) Alpine Valley in morning. See Plate 4–10 for technical details. (Middle) Alpine Valley in afternoon. See Plate 6–4 for technical details. (Bottom) Mt. Wilson and Palomar Observatories, 1957 August 4 04^h 20^m UT, phase 7.88 days, colongitude 10°. Alpine Valley in very early morning.

Any satisfactory hypothesis accounting for these peculiar valley-like features must not neglect the following data:

i. The "gashes" are approximately parallel to each other.

ii. They are parallel to the right and left walls of Ptolemaeus and to the great rill at the right edge of its floor.

iii. They are parallel to the great ridge on the floor of Alphonsus, to a conspicuous rill on this same floor, and to the greater part of a still more prominent rill at the extreme left, Plate 11–2.

iv. The valley-like features are parallel to rills of craterlets on the northern floor of Ptolemaeus.

v. They are rather closely parallel to the Straight Wall, which forms part of the left hand shore of Mare Nubium, Plate 11–1.

vi. They are approximately perpendicular to the slope gradient between Mare Nubium and Mare Tranquillitatis.

(e) *Schiller*. This walled plain (Plate 8–9) is so much elongated that it borders on the valley type of feature. Because of its nearness to the limb, foreshortening distorts it very much. The picture exhibits its rim craterlets strongly and quite definitely indicates that it is a graben-like formation.

Plate 15–7. Oblique view of Alpine Valley by Lunar Orbiter V, with Mare Imbrium on the right. Valley is about 150 km long and 8 km wide, with a sinuous rill down its center. Mode of origin of this combined feature, unique on nearside of the moon, is still a problem.

Plate 15–8. The greatest of the parallel valleys, or "gashes," in the vicinity of Ptolemaeus. See Plate 12–3 for technical details. The valley is straight, except near its middle. Since it runs in the same direction as other linear trends in the terrain, probably it was fault produced.

3. *Messier and W. H. Pickering.* This controversial pair of twin craters illustrates excellently how varying shadows distort features and cause misunderstandings. It has actually been argued that these craters change size and shape during the month. Plate 15–9, from Lick Observatory photographs made at different phases, tells the story rather completely:

(a) A saucer-like depression exists, partly between the two craters.

(b) The southern ray is a series of bright spots; the northern one, under some illuminations, is observed to branch near the middle of its course, with a very faint component turning slightly northward.

(c) The two main rays are tangent not only to W. H. Pickering but also to another crater of approximately the same size farther righ[t]. This crater on the right, especially under a morn[ing] ing sun, is less conspicuous than are the other[s] but only because of the light background. I[n] the afternoon it is more nearly comparable.

The Apollo 11 photograph, which appears i[n] Plate 15–10, helps to clarify the picture. Messie[r,] on the left, is oddly shaped, and seems to b[e] two craters, rather than one simple, round cra[ter.] ter. With this formation, a change in the d[i-] rection of the sun could cause unusual change[s] in light and shade, looking like changes in siz[e] and shape. As for the two rays, a right-left faul[t] line probably exists in the surface rock and en[-] docraters have formed along it, with "blow hole["] craters along the southern ray. If true, thes[e] craters and their rays are no longer mysterie[s.]

Plate 15–9. Messier and W[.] H. Pickering craters and thei[r] ray system, with sunris[e] (above) and morning su[n] (below). See Plates 4–2 and 4–3 for technical details.

Plate 15–10. Messier and W[.] H. Pickering photographed by Apollo 8 astronaut in lunar orbit. W. H. Pickering turns out to be a unique double crater, with an apparent break in the common wall between the old and new craters, and a ray extending from the crater. The shape of the double crater affects the shadows so much that early astronomers believed W. H. Pickering was changing.

6 RESIDUAL OUTGASSING

he resurgence of serious interest in the moon at followed World War II focused attention n the problem of "outgassing," the release of as from the lunar interior. One of the strangest ses involves the crater Linné.

Linné was discovered by Riccioli during the venteenth century; Lohrman, Mädler, and chmidt all observed it as a crater. Schmidt rew it as a crater in eight of eleven drawings e made between 1840 and 1843. In 1866, howver, he made the spectacular announcement at the crater no longer remained and that only e bright mound was observable. During 1867 umerous observers could find only the mound. ate in that year Schmidt announced that he ould discern a mountain in the center of the ound. During 1868 Knott, Buckingham, and ey observed a shallow depression at the center. ater a craterlet was detected by Secchi, who stimated its diameter at barely half a mile. Still ter Huggins measured its diameter as two miles. ince that time Linné has been observed, when ear the terminator, more or less easily through rge telescopes.

In a rather recent program hundreds of pairs f photographs have been made of the lunar sur-

face. The two plates of each pair were exposed with a minimum time interval between them. One of the two plates used only violet and blue light, and the other used infrared light almost exclusively.

The Linné mound has been photographed on some of these dual plates. An infrared exposure is reproduced as Plate 10–15, and its blue-violet mate as Plate 10–16. Both show the surface of the right part of Mare Serenitatis nicely, but the infrared plate is distinctly the better. Because the moon has no appreciable atmosphere, this difference in over-all clarity must be due to the terrestrial atmosphere; the passage of the light through the lower air of the earth absorbs and scatters the shorter violet waves more than it does the longer red ones.

The craterlet on the Linné mound shows sharply in the infrared (Plate 10–15), but is invisible in the blue-violet (Plate 10–16). The difference in Linné's appearance is greater than one would expect from examination of the rest of the plate, although it is not enough to make it absolutely certain that it is not merely an accidental effect from our own atmosphere. If the effect is in fact not due to the terrestrial atmosphere, it must be from something on the moon itself. The reasonable assumption in such a case is the existence of a slight haze between the craterlet and the observer. Such a haze might be either dust or gas, or both. The hypothesis of a dust cloud can no longer be considered favorably, but a very low density of gas leaking from the craterlet would be almost completely ionized by sunlight. The ionization makes it lu-minous and also increases its scattering effe[c]t on light passing through it, especially on t[he] shorter wavelengths.

Indeed, a gas only one-billionth as dense [as] earth's sea-level atmosphere could produce [a] noticeable effect. It appears rather probable, b[ut] not certain, that there is such a residual outga[s]sing from Linné. The fact that such able nin[e]teenth-century astronomers as Schmidt, Secch[i]

Plate 16–1. Mt. Wilson and Palomar Observatories, (left) 1956 October 26^d 12^h 54^m UT, Kodak II–O plate, no filter; (right) two minutes later with Kodak I–N plate and infrared filter; phase 22.35 days, colongitude 177°. Possible obscuration of lower left floor of Alphonsus.

Huggins believed that they had observed
nges makes the evidence for such a conclu-
n much more convincing than if the current
ervations pertained merely to some previously
observed craterlet. If the crater disappeared
described, there was much more of such out-
sing for several months in 1866 than has oc-
red since.

When photographed by Lunar Orbiter (Plate
16–A), no obscuration of the craterpit of
né appeared. Residual outgassing might be a
e event, however, Hyginus rill and crater
late 15–1–B) have also seemed obscured on
asion, as have other features on the moon. All
these obscurations, as well as red flashes and
ches, reported by many observers, are known
"transient lunar phenomena." Recently, some
dence has come in of such events in Alphonsus.

Predawn seeing conditions at Mount Wilson
the morning of October 26, 1956, were un-
ally good. Four pairs of plates were made
t night of the area from Ptolemaeus through
cho. An obscuration of the lower left part
the floor of Alphonsus was evident on the blue-
let plates. Its reality was checked on all four
irs by using Arzachel and other parts of Al-
onsus as controls. Plate 16–1 is one of the four
irs taken on this date. Each observer must de-
le for himself whether there is a significantly
eater loss in the blue light along the northern
rt of the rill on the floor to the left than there
in other places. Such a loss, of course, would
ggest outgassing from this rill, which contains
e famous black spots of Alphonsus with one
more craterlets at the center of each.

Publication of this Alphonsus result led the
viet astronomer Dr. N. A. Kozyrev of the Pul-
vo Observatory to one of the most provocative
servations in modern lunar astronomy. He made
spectrographic study of the region, using the
-inch reflector of the Crimea Observatory with
prism spectrograph. On the night of November
1958, with lunar phase conditions almost the
me as they had been when the haze was ob-
rved in 1956, the slit of his spectrograph was
tended right and left across the central peak
Alphonsus, as shown in Plate 16–2. The lines
the spectrum are images of the slit of the
ectrograph and of a very narrow strip of the
age of any object which has been projected
it. Plate 16–2 shows the position of the slit

as used by Kozyrev for these observations. Each
line of the spectrum actually is a picture of the
narrow strip of the moon which here is hidden by
the ink of the line used to represent it. Each line,
therefore, must get brighter where the moon is
brighter along this strip. Near the bottom of the
slit is the bright, broad image of the wall of
Alphonsus. Therefore the spectrogram must show
a broad, bright band running across all of the
lines of the spectrogram. Our photograph of the
spectrogram is a negative, therefore we see this
band as a dark one. The craterlet which Kozyrev
was observing is projected near the middle of the
slit. To the extent that any spectrogram differs
from that of the sun it must have been affected by
some lunar characteristic. If a significant change
should be temporary it would indicate a temporary action of some sort on the moon. What
Kozyrev reported was a spectrogram that showed
luminescence which must belong to the moon
instead of the sun. A copy of the translation of
his letter to the author is translated for the reader
in the appendix.

Unfortunately the accounts of Dr. Kozyrev's
observations, as reported in our newspapers, suffered both from omission of pertinent data and
from language difficulties. It would have been
far better if the exciting word "eruption" had
not been used. Technically it is correct, but the
image which it brings to mind is not. The "eruption" was not an explosion but a half-hour-long
discharge of very low-pressure gas from a craterlet atop the central mountain of Alphonsus.

The numbers which have been placed over
Kozyrev's spectra are the wavelengths as used
by physicists. If one looks at the top spectrum
(Kozyrev's Number 2) he notices that the negative is darker near the middle of the lines than
it is for the same positions in the other two
spectra. This is the position where, for all three
spectra, the light falls from the central craterlet
of Alphonsus. The darkening of the negative is
especially noticeable for the general section between about 4,000 and 4,800. There is no question but that additional light was coming from
the region of the craterlet when this picture was
made, more than when the ones just before and
after were exposed. The only reasonable explanation is that a low density gas, ionized by the sunlight, was escaping for about half an hour from
this craterlet.

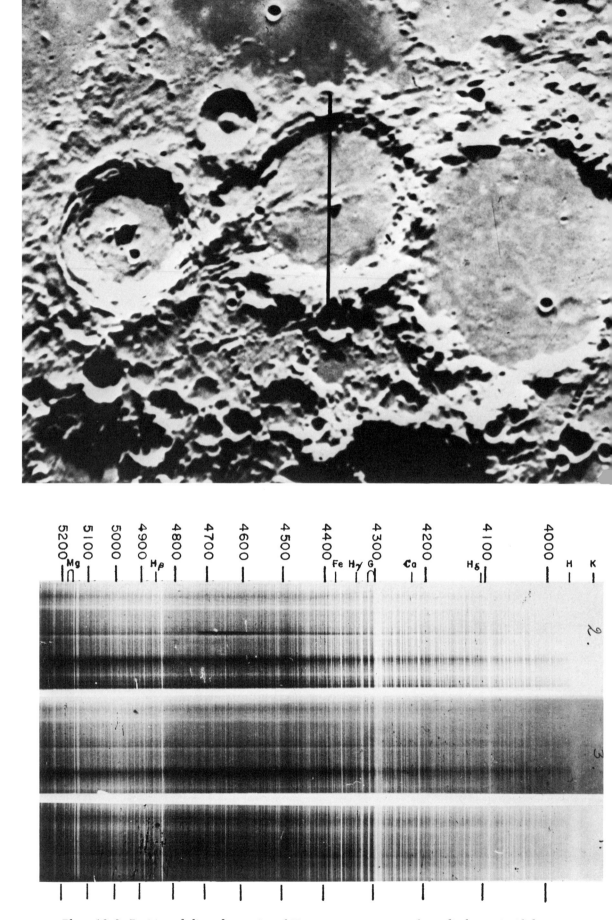

Plate 16–2. Position of slit and negative of Kozyrev spectrograms of gas discharge in Alphonsus.

Although there is no question about the emis-
ı shown in the continuum, nevertheless iden-
:ation of the Swan bands is less certain. These
emission bands from molecules composed
wo atoms of carbon, and are commonly found
cometary spectra where the molecules have
ın excited by solar radiation. Such gases in
:omet must be of almost unbelievably low
ısity. Kozyrev has estimated from his spectro-
.m that the density of the gas at the opening
the craterlet was in the neighborhood of a
lionth of that of our air at sea level. This low
ısity is sufficient to produce the observed effect
1 also that of the direct photographs of October
1956.

Although it would be important to confirm
: Swan spectrum in Dr. Kozyrev's observation,
: fact that an emission spectrum existed, no
.tter what the source, is much more so. With
: possible exception of the Russian photographs
the far side of the moon, Kozyrev's spectrum
the most important single lunar observation
ır made.

When all data from every source are con-
lered, it seems to be almost certain that at
.st some residual outgassing still takes place
the moon. The ringed plain, Plato, is another
.ce where evidence of residual outgassing has
ın reported. Rarely, observers have been un-
le to see the many little craters on the floor of
ıto, photographed by Lunar Orbiter in Plate
-3, although the seeing is good enough to reveal
ım. Perhaps there is quite a lot of activity of
s kind, for over the last four centuries about
) obscurations, pink patches, and bright red
ıws and flashes have been reported on the moon.
On October 29, 1963, an American lunar map-
ı, James Greenacre, saw pink patches in three
ots around Aristarchus (Plate 16–4). The spots
ninded him of the neon lights in which red and
ıite streaks chase each other around the edges
signs. Two others with him saw the flashes
rough the 24-inch Lowell Observatory telescope
Flagstaff, Arizona, and another observer at the
arby Perkins telescope confirmed the lights.

The bright spots and hazy patches have been
ported most often in fresh, "young" features, like
·istarchus, Tycho, and Alphonsus. Points around
e rims of lunar maria, like Mare Crisium, Im-
ıium, and Serenitatis also have been seen to
ow and flash with transient lunar phenomena.

Plate 16–3. The mare-filled Plato, photographed by
Lunar Orbiter IV from 2886 km. Plato is about 100 km
in diameter. Lighter material patches its floor and
sinuous rills are near its rim.

Plate 16–4. Aristarchus and vicinity from Lunar Orbiter IV at an altitude of 2668 km. Herodotus lies to
its right and Schröter's Valley below, with craterlet
chains. See closeup in Plate 21–4.

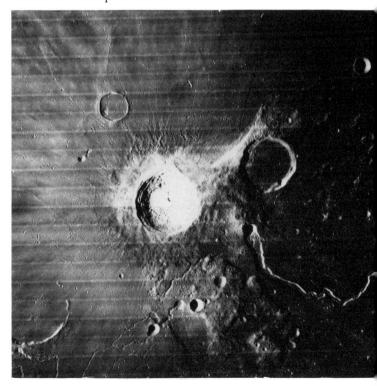

Plate 16–5. Three Thebit craters photographed by Lunar Orbiter IV at altitude of 2719 km. Ridges and clefts appear within the largest crater, which has slumping walls. Compare with Plate 16–6.

Plate 16–6. Model similar to Thebit formation, made by fluidization of powdered rock by jets of air from below, in experiments by A. A. Mills, University of Leicester, England. Compare Plate 16–5.

Early astronomers thought that the lu... transient phenomena might be volcanic erupti... triggered perhaps by earth-caused tides on ... moon. Since the moon causes sizable tides on ... earth, think how much stronger the tides ... earth causes on the moon must be. (See Chap... 17). The quiet, unchanging character of the m... convinced later astronomers that the moon v... past the stage of actual eruptions in which mas... of lava and clouds of dust and gases would po... out.

Many other theories have been proposed to ... plain the transient phenomena. Could they be ... flections from bright-colored crystals in lu... minerals on the surface, flashing when the s... reached a proper angle? Or are there materi... that luminesce, or glow, when showered with t... nuclear particles in the solar wind, or ultravio... light in the sun's rays? With no detectable atm... phere and without a strong magnetosphere li... the earth's the lunar surface is exposed to t... full furies of the sun.

Recently, British scientists bombarded ve... cold granite, as well as meteorites and natu... glasses, with nuclear particles, protons, and th... heated the materials. As they warmed, the ma... rials, particularly the granite, but also the met... orites and the glasses, glowed intensely red, wi... thermoluminescence. This might be the expl... nation for the red glows on the moon, the scie... tists suggested. After being bombarded by pr... tons in the solar wind for half a million yea... these materials on the moon might give off lig... of the same order as reflected sunlight. When t... materials were uncovered by slides or meteorit... impacts, or even minor earthquakes on the moc... their warming in the moon's next day could mal... them luminesce brightly.

Another theory has recently been propose... by the British scientist A. A. Mills, to explain t... way some lunar craters look, as well as lun... transient events. This suggestion is based on t... process of fluidization, used in many industri... in which finely ground particles, like coal du... are blown through pipes by air currents. Mil... set up an experimental plastic box, partly fille... it with powdered rock, and blew jets of air ... through the powder. The craters that formed ... the rock dust looked like some lunar craters.

Compare Plate 16–5, the triple lunar crat... Thebit, with Plate 16–6, one of the model form... tions produced by fluidization in powdered roc... The resemblances are striking. Plate 16–7 shov...

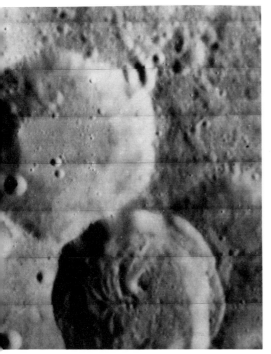

e 16–7. Twin craters of Abenezra (below) and phi (above), photographed by Lunar Orbiter IV, he south-central lunar highlands. A fairly straight, l wall divides the craters.

Plate 16–8. Twin fluidization craters similar to lunar craters Abenezra and Azophi (compare Plate 16–7), produced by air jets up through rock dust by A. A. Mills, University of Leicester, England.

twin lunar craters Azophi and Abenezra, as tographed by Lunar Orbiter; the comparison h the craters produced in a fluidized rock bed Plate 16–8 again is intriguing. Plate 19–21 ws another pair of craters on the farside of the on, which may have been shaped by fluidiza-.

Mills has also proposed that the lunar tran- nt events may be caused by fluidization. Even h gentle outgassing from below the surface of moon, enough dust might be raised and car- d to create electrostatic glow discharges from ositely charged regions within the clouds, ch as lightning displays are to be seen within billowing ash clouds accompanying volcanic ptions. On the moon, the predominance of lrogen gas in the dust might give the reddish t to the transient phenomena.

The moon has been much more frequently served from earth in late years, in coordination h the space program. As observing increases, frequency of reported transient events has o increased. The still unsettled problem of tgassing shows the need for systematic obser- ion of the moon from earth. Observations de simultaneously over the whole of the moon's arside can be secured only from the earth.

There is no truly satisfactory scale of bright- ness for lunar features. The present obsolescent scale, described in Chapter 2, should be re- placed. Accurate knowledge of relative bright- nesses is essential to any program involving study of possible lunar changes. A photometric pro- gram also has much value in studies of the origin of the surface features. Approximately a hundred of the most important areas should be measured by use of modern photometric equipment, at small intervals of colongitude, throughout the month and the variations of albedo should be catalogued for each. Corrections for librations should be included in the case of all features near the limb.

All intensely black and intensely bright cra- terlets should be investigated spectrographically throughout each month, as should such ringed plains as Alphonsus, Aristarchus, and Plato. Only through such a comprehensive program does there appear to be much chance of gaining knowledge concerning sites of outgassing and of the frequencies of its occurrence. Such a pro- gram should also locate the principal conditions of luminescence that may exist in craters, rays, scarps, and other topographic phenomena.

17 TIDES TODAY AND
THROUGH THE AGES

The friction of the tides pulling against the surface of the earth slows the earth's rotation, and this in turn gradually affects the apparent motion of the moon around the earth. The tidal forces also produce, although with almost infinite slowness, tremendous changes in the moon's actual orbit. Indeed, all the ordinary perturbations to which that orbit is subject, except the one due to the equatorial bulge of the earth, are attributable to forces which we might consider tidal.

An elementary explanation of tides is easy enough, but a thorough study of them is the work of a lifetime. A rather simple tidal theory was worked out by the great French astronomer and mathematician, Laplace. George Darwin, the astronomer son of Charles Darwin, contributed four large volumes of research and summarized them in one much smaller and less technical book. Of course long before Laplace the great Newton saw that the tides were implicit in his laws, and gave a first simple explanation of them that tells enough for some purposes.

Contrary to what one might expect, tides

tend to be high when the moon is at its highest and also when it is 180 degrees away, high over the opposite side of the earth. The tide is low when the moon is near the horizon. If the ocean were deep enough and had smooth floors with no islands or continents, and if the earth rotated slowly, we would experience the highest tide almost exactly when the moon is nearest the zenith and when it is 180 degrees from the zenith, and low tide with the moon on the horizon. The lack of these ideal conditions is responsible for the lags that produce most of the complications.

What seems surprising is that there should be a high tide when the moon is farthest from, as well as at, the zenith. This is because we tend to think of the earth as a hard ball with a rubber skin of oceans around it, and we think of the moon as reaching out hands to grasp and to pull on it. The rubber skin, by this analogy, would be pulled out somewhat to produce a high tide on the side toward the moon and would be stretched tight, or thin, on the opposite side, producing a low tide there.

But all of this thinking is incorrect. As New-

stated, "every particle attracts every other
rticle . . .", so each particle of the moon is
lling each particle of the earth. It does not
tter a bit whether the particles are in our air,
r oceans, or our rivers, on the "solid" surface
deep inside the core of the earth. Each one
them is pulled by each particle anywhere
hin the moon. In turn each terrestrial particle
lls each lunar particle just as strongly.

If this were all there were to the law there
uld be no tides, no matter how strongly the
on pulled on the earth. The tides result from
relation Newton expressed in the words that
low immediately upon his phrase quoted
ove: ". . . inversely as the square of the dis-
ce."

In Figure 17, A and B are two equal particles,
e in the ocean, the other near it in the "solid"
e of the earth. C is any particle of the same
ss near the center of the earth, about 4,000
es farther from the moon than are A and B.
and E are near each other and about 4,000
es still farther from the moon than the earth's
ter. The moon pulls A and B by the same
ount, and so they tend to "fall" the same dis-
ce toward the moon each second. The moon
lls on C less strongly because C is 4,000 miles
ther away, so C does not tend to fall as far
the first particles. Consequently, the distance
increases, and we have a high tide at A.
ce D and E are 4,000 miles farther than C
n the moon, they tend to fall less than does
and as a result we have a high tide at E also;
center of the earth is pulled away from the
er at E. The low tides, then, must be on a
at circle around the earth and half way be-
en A and E.

If the whole of the earth were fluid, like
oceans, it would be very difficult to detect
tides. We would have no ocean bottom to
us. Of course this is not the case; most of
earth is part of what can be considered, in
context, a rigid sphere. B, C, and D are tied
ether by the earth's rigidity and must fall
ard the moon by nearly the same amount,
n though they are pulled differently by the
llite. The average of the effects on B and D
als the amount C tends to fall. But because
nd E are both fairly free from other particles
he ocean, they fall toward the moon by ap-
ximately the amounts demanded by their dis-
es. As a result the ocean must get deeper at

A and at E. A moves a bit away from B because
B is tied to C, and D moves away from E because
D is tied to C. Actually the solid part of the
earth is not perfectly rigid, and the distances
CB and CD do change by about six inches twice
a day. Our houses rise and fall in this manner all
the time.

This explanation applies to points A and E,
where the moon is passing through the zenith.
These points are always in or near the tropics
of the earth. In areas near the poles where the
moon cannot pass through the zenith, tidal forces
are never as great as they can be in the tropics.
At the poles the moon must always be low in
the sky and when it is at the celestial equator,
twice a month, there must theoretically be a
continuous low tide at the poles all day long.

The tidal force, then, is merely the difference
between two forces. One of these is the moon's
attraction on a particle at the center of the
earth, the other its attraction on a particle at the
point on the surface where we are studying the
tide. It can be shown from Newton's law, using
only college freshman algebra, that the tidal force
varies approximately as the inverse cube of the
moon's distance from the earth. If the moon were
twice as close, the force would be eight times
as great. If it were twice as far away, it would
be one eighth what it is. During each month the
moon's distance varies between 221,000 and
253,000 miles, and of course this is enough to
cause the very noticeable monthly variation in
tidal force.

The sun pulls the earth about 160 times as
strongly as does the moon, but it is 400 times as
far away and so its tidal force is only about two
fifths as great. In other words the difference between the sun's pull on A and on C is only two
fifths of the difference between the moon's pull

Figure 17. Tides caused by the moon.

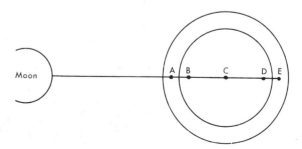

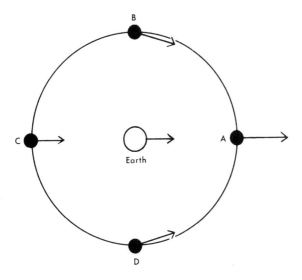

Figure 18. Effect of the sun on moon's orbit.

moon's distance from the earth varies duri[ng] the month, the sun is in the plane of the moo[n] orbit at only two instants during the month, a[nd] the sun's distance from the earth changes duri[ng] the year. These three complications cause ma[ny] periodic terms affecting the orbit of the moo[n] In addition to the sun each one of the plane[ts] is pulling on the earth and moon and creates very slight perturbation. E. W. Brown in [his] great study of the moon's orbit included m[ore] than 650 such terms in his lunar tables.

The two most important of these many p[er]turbations affect our calculation of eclipses ve[ry] much. In Figure 19, showing the elements [of] the moon's orbit, the line FB, along which t[he] plane of the earth's orbit around the sun and t[he] plane of the moon's orbit around the earth int[er]sect each other, is called the *line of nodes*. T[he] line AP, which joins the two points P and [A] where the moon is respectively closest and fa[r]thest from the earth during each month, is call[ed] the *line of apsides*. If there were no perturb[a]tions, these lines would always point in the sa[me] direction in space. But the perturbations [do] exist, and because of them these lines revol[ve.] The line of apsides turns forward at an irregu[lar] rate in the plane of the moon's orbit, requiri[ng] on the average a trifle less than nine yea[rs] (8.8503) to go around once and return to [its] original position. The line of nodes turns ba[ck]ward in the ecliptic (the plane of the eart[h's] orbit), reaching its original position in 18.5[9] years. An eclipse takes place only when the mo[on] is close enough to one of its nodes at the insta[nt] of full or of new moon. Moreover, a total so[lar] eclipse can occur only when the moon is not [too] far from perigee, the point in its orbit where [it] is closest to the earth. When it is far from perig[ee] its angular diameter is less than that of the s[un] and hence it cannot completely obscure the s[un.] The effects of these perturbations on the pred[ic]tion of eclipses are described simply in Chapt[er] 20 and 21 of *Pictorial Astronomy* (Thomas [Y.] Crowell Co.).

The tides cause constant friction between [the] ocean waters and the solid portion of the ea[rth.] Because the month is longer than the day, t[his] friction slows down the earth's rotation. If [the] month were shorter than the day the frict[ion] would speed up the rotation. If the day a[nd] month were equal the areas of highest tide wo[uld]

on those points when both the sun and moon are at the zenith. The inverse cube law cuts down the tidal force very rapidly.

All of this is very much complicated in practice by a variety of other effects: the combinations of lunar and solar tides, the shallowness of oceans, blocking or obstructing of tides by islands and continents, and the shapes of shorelines and other terrestrial factors.

The sun's perturbing force on the moon's orbit, which is actually another tidal force, with the earth and the moon as the particles affected, is illustrated in Figure 18, in which the moon is of course much too close to the earth. When the moon is at A, the sun pulls more strongly on each particle of the moon than it does on similar particles of the earth. Therefore, it pulls the moon away from the earth. When the moon is at C the sun pulls the earth away from the moon. When the moon is at either B or D the pull of the sun on each particle of it is about as strong as the pull on the earth's particles, but the directions of the attractions toward the sun's center converge in the direction of the sun's pull on the earth. As a result, the sun tends to pull the earth and moon a little closer together at such times. This effect, evident near the half moon phase, is like the low-tide effect referred to in earlier parts of this same chapter.

The perturbations of the moon's orbit by the sun are complicated by several other factors. The

ot change on the earth and there would be o friction. Two thousand years ago a day was bout $\frac{1}{50}$ second shorter than it is now. So the verage day of the last 2,000 years was $\frac{1}{100}$ econd shorter than a day is now. Multiplying $\frac{1}{100} \times 365 \times 2000$ we find that 2,000 years ago he position of any point on earth, in terms of the lobe's rotation, was displaced by 7,300 seconds f time, or about two hours, from where it would ave been if the length of day had been constant. he displacement 4,000 years ago was about ight hours. This displacement affects the times f occurrence of eclipses of the sun and their po- itions on the surface of the earth to such an xtent that even some of the imperfect ancient clipse records can be used to check the change i rotation.

The effect of the tides in slowing down the arth's rotation decreases the energy of the rota- on and involves us with another universal law f physics, one that is as important as the law f gravitation. This is the law of conservation f energy: Energy can neither be created nor estroyed.

Energy can, of course, appear in many forms nd can be changed from one form to another. he most condensed form by far is as matter.

(We even know the exact relationship between grams of matter and ergs of energy.) Energy can also appear as motion (kinetic energy), as energy of position (potential energy), as work done, as electricity or heat, or in still other forms. The amount of work done on a body measures the change in its energy. Since energy cannot be created or destroyed, if energy is taken from one place, as it is when the tides slow the rota- tion of the earth, it must automatically appear in some other place. A little of the lost energy of rotation becomes heat energy and is radiated into space. The remainder must stay somewhere in the earth-moon system. When it is tracked down mathematically it is discovered to be at work pushing the earth and moon farther and farther apart! This change in the moon's distance is very slow. But if humanity can maintain its existence and its interest long enough to observe the development, the moon will one day be seen twice as far away as it is now and the tidal force will be only an eighth as strong. Once, far away in the past, the moon must have been at half its present distance, exerting tidal forces eight times what they are now upon an earth spinning far more rapidly than it does today. But this is part of the story of the evolution of the moon.

Figure 19. The orbit of the moon in respect to the ecliptic.

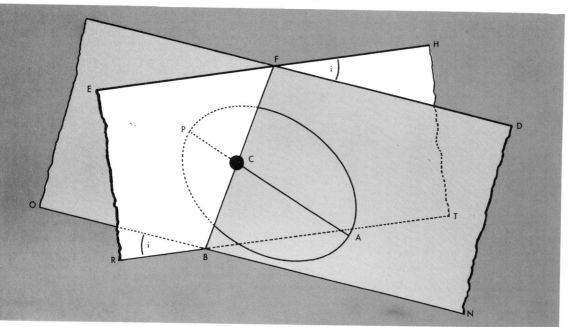

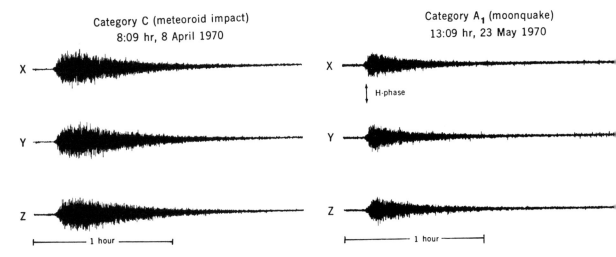

Category C (meteoroid impact)
8:09 hr, 8 April 1970

Category A₁ (moonquake)
13:09 hr, 23 May 1970

Figure 20. Records received on earth from two seismic events on the moon, the impact of a meteoroid (left) and a moonquake (right). X and Y show horizontal motion in two perpendicular directions, Z shows vertical motion. The H-phase is larger for moonquakes than for lunar impacts. (From "Moonquakes," G. Latham *et al., Science*, Vol. 174, No. 4010, Figure 1.)

The dominant role of earth's tidal forces on the moon has been revealed in the first seismograms telemetered to earth from instruments left on the moon by Apollo astronauts. Hundreds of shaking events have been reported, from both the impact of meteoroids on the surface and of moonquakes deep under it. Two of these are shown in Figure 20. Most of the moonquakes occur near the time of perigee, when the moon is closest to the earth and its tidal attractions are strongest.

The moonquakes have been concentrated along two focal zones running in great circles from north to south and northeast to southwest across the nearside of the moon. The quakes appear to come from sources of strain deep within the moon, for they originate up to several hundred miles below the surface. Still the moonquakes are so slight that they would hardly be noticed by an astronaut standing on the surface, and the meteoroid impacts have actually been stronger. Swarms of moonquakes have been observed, when quakes come as often as every two hours for periods of several days.

The reality of moonquakes reverses the long-held idea that nothing ever happens on the moon. It may be that vibrations of this kind within the moon, released by tidal stresses, bring about the residual outgassing referred to in Chapter 16. The vibrations of moonquakes last for hours, longer than earthquakes of similar intensity.

While the moon has been shaken by quakes and the impact of meteoroids, beams from radio telescopes have also been impinging on its surface reflecting back to earth. This radar technique has now been developed to the point at which, with wavelengths of less than two inches, very precise

Figure 21. Altitude profile around the moon made by the laser altimeter of Apollo 16. Variations (ΔR) from the radius of a spherical moon are shown in kilometers (km). One km equals 0.62 miles. Only lunar longitude is shown, though the profile runs at an angle to the equator.

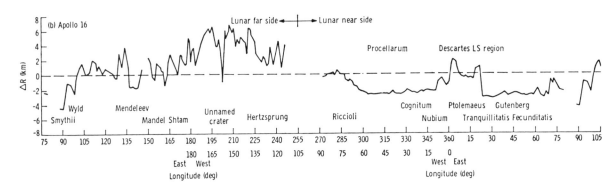

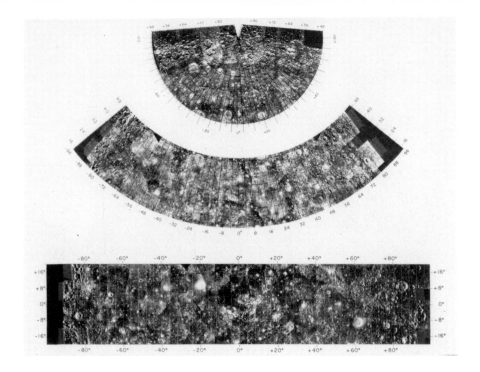

Plate 17–1. Map by radar reflectivity of part of the moon. Each rectangle is about 2 km on a side of the lunar surface. Lunar latitude and longitude are given. Alphonsus appears just below and to the left of 0° longitude. (See Lunar Earthside Chart, Chapter 19.)

maps of the lunar surface can be produced, like that in Plate 17–1. These clearly reveal small variations in lunar topography, as well as in the roughness of the surface. Soon, using a technique of two radio antennas interferometrically in connection with radar reflectivity, more accurate estimates of lunar surface heights over wide areas may be made.

Apollo 15, 16, and 17 astronauts have now obtained the most accurate measurements of a series of lunar altitudes ever made, using the reflection of a laser beam from the moon every 20 seconds as the command module orbited. the moon. Figure 21 shows the profile of lunar altitudes measured by Apollo 16. The laser beam had a spot size on the lunar surface of 100 feet and gave an altitude accuracy of six feet!

According to these laser altitudes, the nearside bulge of the moon toward the earth is no more than 2 km (a tenth of some prior estimates), so the moon is more nearly spherical than was believed. Many altitudes given by the laser disagree with those on lunar maps, in which major revisions will have to be made. Thus, Ptolemaeus now proves to be half a mile higher than Mare Nubium, not lower as shown on existing maps.

Perhaps more significant, the laser showed that the moon's center of gravity is offset about half a mile toward the east of the optical center as seen from the earth, and with the bulge, offset about 2 km toward the earth. What could produce such an offset in a moon that is so nearly spherical in shape?

One answer to this question might be found in the lunar mascons, or mass concentrations, that have been discovered since space vehicles were placed in orbit of the moon. Irregularities in the paths of Lunar Orbiters and Apollo vehicles showed that certain places on the moon had stronger gravity than others. Concentrations of mass in these areas would increase the gravitation. Mascon areas were soon pinpointed in such maria as Imbrium, Serenitatis, Crisium, Marginis, and Orientale. Another is suspected on the mid-farside of the moon, called "Occultum." It is speculated that the bodies, like asteroids, that impacted to trigger the formation of these maria, may have settled in dense mass concentrations under their lava surfaces.

The mascons are too small, however, to account for the whole offset of the moon's center of gravity. Perhaps, in addition, there are large negative mass areas on the moon's farside. More likely, the core of the moon may be offset toward the earth, with a thinner crust, and, therefore, more and larger maria, on the nearside.

18 THE EVOLUTION OF THE MOON

Our moon is a freak among the satellites of the solar system. It is true that Jupiter and Saturn do have some moons which are larger than our moon but those planets are huge. When we consider ratios of masses of satellites to their primaries none of them is even comparable to that of our moon. If there were astronomers on other planets we can be certain that their name for us would be "The Double Planet." It is a freak in a second manner. The two satellites of Mars move almost exactly in the plane of that planet's equator. One of them diverges by half a degree, the other by only a thirtieth of that tiny amount. Four of the five inner satellites of Jupiter are so nearly in the plane of that body's equator that the difference is not measurable. The fifth diverges by less than half a degree. The seven satellites which are closest to Saturn also have orbits which lie practically in the plane of Saturn's equator. The third satellite has the greatest inclination and that is merely a trifling 66 minutes of arc. The case of Uranus is even more spectacular. The orbit plane and the planet's equator are almost perpendicular to each other. The five satellites move in the equatorial plane, insofar as we are able to determine. Our moon's orbit around the earth lies rather close to the earth's orbit around the sun. The angle between the plane of the moon's orbit and the plane of the earth's equator varies between 18½ and 28½ degrees in a period of about 18.6 years.

In shape of its orbit the moon also exhibits the same sort of freakishness. Its orbit is quite eccentric, the distance from the earth varying between 221,000 and 253,000 miles during the month. Neither of the satellites of Mars has even half this eccentric an orbit. All five of the inner satellites of Jupiter have orbits which are practically circular. The outer seven, which probably are captured asteroids, all have orbits with greater eccentricities. The six inner satellites of Saturn all have orbits which are less eccentric than that of our moon and only the sixth has even half the eccentricity. The satellites of Uranus all have orbits which differ only negligibly from circularity.

Much about the evolution of the moon can be deduced from the physical laws governing tides, which were discussed in the previous chapter. In that chapter we considered briefly the law of conservation of energy. This is the law which makes it impossible in the physical universe to get something for nothing. It is the law

hich prevents our everyday lives from under-
ing the freakishness of nightmares. Closely re-
ted to energy is *momentum*, a useful quantity
hich possibly we could consider as having been
vented by man. It is defined as the product
the mass (amount of material in a body) mul-
lied by its velocity.

Directly related to momentum is *moment of
omentum*. For a revolving object, this is the
omentum of the object multiplied by the radius
its orbit. Momentum itself is not conserved the
ay energy is, but moment of momentum is con-
rved.

A familiar example from everyday life il-
strates the conservation of moment of momen-
m. Figure skaters often twirl on the point of
e skate. The skater usually starts slowly, with
ms outstretched. This is even more spectacular
he holds a weight in each hand. Then he
ings his arms close to his body—and spins
ster. The increased speed results from conserva-
on of moment of momentum: he has reduced
is radius of rotation and to keep the moment
nstant, his velocity must increase. If he raises
is arms, he will spin more slowly again.

The same situation applies to the earth and
oon. They form a rotating, two-body system.
he sum of their moments of momentum must
emain constant except for the fact that small
mounts of energy are lost to space. Thus the
ss in moment of momentum of the earth, due
tidal friction, must be equaled by the gain in
he moon's moment of momentum. The equations
re too complicated to justify inclusion here. Momentum is defined as the product of mass times
velocity. The equations describing energy and
momentum will not be used in this discussion,
ut they do make the relationships clear: when
represents energy of motion (kinetic energy),
stands for momentum, m is the mass and v
he velocity of a body, then $M = mv$ and $E = \frac{1}{2}mv^2 = \frac{1}{2}Mv$.

The lunar orbital moment of momentum is
t present almost five times that of the earth's
otation, but the sum of the moon's and the
arth's moments must remain essentially con-
tant. From these relationships we can tell how
uch the month will lengthen for any given
hange in the length of the day. It has been
etermined that although both the day and the
onth are increasing, the day is lengthening

more rapidly. This produces a tendency for the
day and the month eventually to become equal.
It follows that if the changes continue long
enough the earth will eventually keep one face
toward the moon at all times. The tides have
already produced this condition on our much
smaller companion. Such an equality would
come when the day and month have both become about 47 of our present days long.

The word "tendency" was used deliberately
in the preceding paragraph. If the oceans were
to freeze, the tidal friction would decrease to a
rather small fraction of what it is now. This
would very much lessen the rate of change and
increase the time necessary to produce the final
effect. Under our present view of stellar evolution it appears very unlikely that such freezing
will ever take place. Still, the whole process is
so slow that the earth and moon may not last
long enough to attain this equality of day and
month.

As the day approaches the month in length,
the effect of the lunar tides in lengthening both
periods must decrease. If they were to become
equal the tidal effect would vanish, for there
would then be fixed tides on both bodies and

Figure 22. Relative size of satellites. The
largest satellites of earth, Jupiter, Saturn,
Uranus, and Neptune are shown. The large
circle represents successively each of these
planets as compared to its largest satellite.

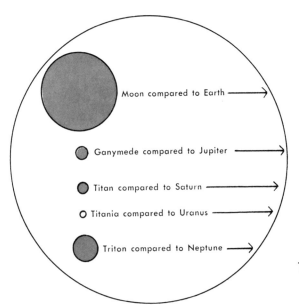

therefore no friction. If the earth and moon could then be left alone the system would be stable—at least we do not know today of anything that would change it from within itself. But all of the foregoing conclusions have ignored the fact that the sun is in the picture too. Its tidal forces on the earth and moon will also have decreased, but at a far lesser rate. The friction of those solar tidal forces on the earth cannot become zero unless the day and the year become equal. With a decrease in the lunar tidal effect the solar tides will eventually become by far the more important tidal factor. They slow down the day without having a long-term effect on the distance between the earth and the moon. This speeds up the time when the day and the month become equal, and if at that point in time the solar tides were to cease, nothing more would happen.

However, the solar tides have a far longer course to run than do their lunar counterparts. They continue to lengthen the day with almost no effect on the month, so the day must eventually become the longer of the two periods. When this happens, the lunar tidal effect must immediately come into existence again. However, now the friction must drag ahead instead of backwards. The conservation of moment of momentum will draw the moon in and thus shorten the month. Nevertheless, the day will remain always the longer. Eventually (if the earth and moon last long enough) the day and the month will both be much shorter than they are now. The moon will be much closer than it is and the tides will be tremendous. When the moon's distance has diminished to one fifth today's value the tidal force will be 125 times as great as it is today.

All this has led some writers (not scientists) to predict that the moon will crash into the earth. Indeed some enthusiasts apparently expect this to happen while man still lives here! They forget, or else never knew, that inverse cube factor of tidal forces. At some future time while the surfaces of the earth and moon are on the average still about a thousand miles apart, the earth's tidal force on the moon will become greater than the gravitational force that holds the moon together. Astronomers call the distance at which this occurs "Roche's limit." At this distance disruption of the moon must begin. Unless the moon's own substance produces very violent explosions, the tidal force will disrupt

the lunar surface, continually working its disruption deeper and deeper. We do not know whether the catastrophe will come as a sudden explosion or as a great cracking of the lunar surface that for a time leaves the primary mass little affected. Gradually, however, all the material which now composes the moon will separate to form a ring around the earth somewhat similar to one of the rings of Saturn.

The earth's greater gravitational force would prevent its being shattered at the time of the moon's destruction, although the immense tides would probably change its whole surface permanently. During the time of destruction the scattering of the lunar fragments would very rapidly lessen the tidal force. Then the system would become quite stable, except insofar as the solar tides increase the distance from the sun, with a consequent gradual lengthening of both the day and the year. It must be remembered constantly, however, that because of the tremendous time factor involved we do not expect these later events ever actually to take place.

Some people have suggested that the rings of Saturn were formed in this manner by a number of satellites approaching too close to a primary. But it seems to be certain that the particles composing these rings were always close to Saturn and that they never were united as one or more normal moons. One of the moons of Mars is even now on its way in to the planet. It is so small, though, that tidal forces are unlikely ever to shatter it.

George Darwin used this tidal process as a basis of his suggestion that the moon evolved from the primitive gaseous earth. His hypothesis was regarded with much favor for a good many years, but today, despite its apparent plausibility, it probably has little except historical importance.

Darwin carried his tidal calculations backward through unknown ages to arrive at an era when the day and month were equal and the earth and moon were very close together. (He made the implicit assumption that the earth-moon system had existed for a long enough time.) He could not state the length of that original day-month more accurately than between three and five of our present hours. With earth and moon very close together, he believed, the tremendous tides must have formed a bridge connecting the two bodies. With day and month equal there would be no tidal friction and, un-

ss there were interference from outside forces,
e configuration would have continued indefi-
tely. However, the resulting equilibrium would
ave been very unstable, quickly upset by the
lar tides. As soon as any inequality developed
ere would be tremendous tidal friction that
ould begin the moving-apart process which he
d followed in reverse.

To account for the nature of the earth and
oon prior to the era reached by his calculation
ackward in time, Darwin had two possible
arting points. He found that if the earth had
en a highly compressible gas, with at most
rather small, dense core, one evolution process
d been followed. If the earth had been a fairly
compressible fluid or even a gas that was not
ghly compressible, the development had been
ong different lines, as shown in Figure 23.

Through theoretical hydrodynamic calcula-
ns the French mathematician Henri Poincaré
d studied the effects of rapid rotation on any
quid mass other than a highly compressible
s. He found that if the rotation were rapid
ough the body would be distorted into a pear-
aped object. The lower part of Figure 21,
hich does not pretend to quantitative accuracy,
ows qualitatively the variations in shape to
hich such a rotating body is subject. Each
ure exhibits the effects of more rapid rotations
an does the one to its left. Darwin showed that
any additional force caused the smaller mem-
r of such a pair to revolve about the larger
rt more slowly than the latter rotated, tre-
endous tidal friction would result immediately.
e neck of the hourglass would be severed with
most explosive rapidity and a satellite would
launched on a career similar to the one we
ve considered for the moon. In this situation
e moon need not be of negligible size as com-
red to its planet, as in the case of the equatorial
g process. Thus no objection arose from the
t that our moon is far larger in comparison
its primary than is any other satellite in the
lar system. Darwin concluded that the birth
our moon probably occurred in this manner
d the astronomers of that time generally agreed
ongly with him.

But apparently the earth never had the neces-
y moment of momentum to have carried the
ar-shaped figure to the stage where the neck
s constricted to a narrow bridge. Working
ckwards from today brought about equality

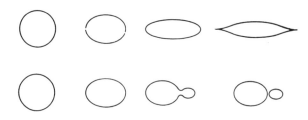

Figure 23. Shapes of rotating fluids. Figures
of equilibrium of gaseous bodies change with
the speed of rotation (increasing speed to
the right). Top row: near-perfect gases; bottom row: viscous fluids.

of day and month while the two bodies still were
too far apart for tides to form the bridge. Darwin attempted the addition of what he called
"tidal resonance" to bring about this stage where
solar tides have separated the bodies. Later
work by others showed that his additional effect,
although present, was insufficient.

A moon formed in this manner through hourglass stages should move in the plane of the
planet's equator just as a moon which evolved
from the ring of equatorial particles. Darwin
attempted to show from theoretical conditions
how the present large angle between the planes
of the moon's orbit and the equator of the earth
might develop eventually.

The recognition of the lack of moment of
momentum stimulated search for a substitute for
Darwin's hypothesis. In 1951, G. P. Kuiper
brought out a new form of evolutionary hypothesis for the planets of the solar system. He
postulated centers of condensation which he
called *protoplanets* in a flattened, rotating nebula with an early stage of the sun as center.
The masses were larger than are the corresponding planetary masses today. If in one of the
protoplanets there were two centers of condensation a satellite could form. Its mass might
be large or small and it would not necessarily
rotate in the plane of the planet's equator although there would be a tendency for that to
be true.

Kuiper's hypothesis appears to most of us as
the most promising which has been offered for
the evolution of planets. We can guess that de-
partures from it will consider only details. Also
it seems to account well for evolution of satellites. However, our moon system is a freak and a

hypothesis which fits the facts other places may not be the one which applies here. The main difficulty comes from the shape of the moon itself, as expressed by the *moments of inertia of its figure*. No satellite, unless it were surprisingly rigid, could be a perfect sphere. The tides of the primary must warp its shape and the nearer it be to that primary the greater must be the departure from sphericity. Our moon does not have exactly the shape that is demanded by tidal theory. It has been assumed that long ago it became rigid when the tides were greater than are those today and that since that time it has not kept pace with their changes. However, calculations have been made which indicate that there may never have been a time when it did conform. If this be true there can be only two possible explanations. The one is that there is a peculiar distribution of mass within it; the other that it has been a rigid body throughout the whole time that it and the earth have formed a system. Urey has attempted to meet the question by a very ingenious assumption that the surface has been loaded by meteoritic accretions of dense material since the time that it became rigid. His explanation well may be the true one. Unless it is accepted we must go to the always rigid moon. Such a body would be impossible under either Darwin's or Kuiper's hypothesis. The initially rigid moon would involve capture by the earth of an independent, small planet. The idea of such capture is a very old one but has been neglected since Darwin's time. The general idea has been worked out recently and appears to predict the general convulsive force needed to create the lunar maria, to explain their peculiar distribution, and to account for the lack of them on farside.

With Lunar Orbiters reporting data about the moon, as well as Apollo astronauts leaving instruments on it and bringing back rocks, some headway has been made on the evolution of the moon. Presumably, both earth and moon began to form about 4.6 billion years ago with the rest of the solar system. Lunar rocks as old as 4 billion years, older than any yet found on earth, have been returned. For about the next billion years, the moon continued to evolve, according to the record in the rocks. Early in this period, it was probably largely or partly molten, so that the lighter minerals rose to its surface, and chemical differentiation took place; then a solid crust formed relatively rapidly.

From about 3.7 to 3 billion years ago, t[he] great maria were formed, triggered by impac[t] of meteoroids. For a long time, during this p[e]riod, lavas welled across the crust, forming t[he] maria as we see them now. This made the man[y] layered crust, found, for example, by the Apol[lo] 15 astronauts in the rim of Hadley Rill and visib[le] in strata on Mount Hadley. New impacts forme[d] the "younger" craters much as we see them no[w].

Surveyor soft-landing vehicles and Apollo a[s]tronauts found a layer of dust, pebbles, a[nd] broken rocks on the lunar surface, quite simi[lar] to the soil in a plowed terrestrial field. T[he] seismic vibrations reported from the moon th[en] enabled scientists to formulate some idea of t[he] lunar crust. The speed of the seismic wav[es] through the crust indicated churned rock fra[g]ments mixed with soil for perhaps the first 3 mi[les] below the surface. Below a few miles to 15 mi[les] in depth, a series of layers of basaltic rock a[p]pears, probably from lava flows. From 15 to [25?] miles, there is solid rock, mostly anorthosites, fr[om] which the highlands of the moon also are co[m]posed. Below this, the waves travel much fast[er] in what may be the lunar mantle. Whether or n[ot] the mantle is near the melting point and the c[ore] is molten are still-unresolved problems.

The laser altimeter readings from the Apo[llo] command modules gave precise lunar altitud[es]. These indicated an offset lunar center of gravi[ty]. This, as well as the lunar mascons and the mar[ia] are explained, some scientists now believe, by [an] offset core within the moon. If the core is abo[ut] 875 miles in diameter, and is offset toward t[he] earth on the nearside of the moon, the man[tle] and crust on this side may be only about 3[?] miles thick, as opposed to 925 miles thick on t[he] farside. With the core so much nearer the surfa[ce] on the nearside, molten lava might more eas[ily] break through to the surface to form the mar[ia]. As lava from the depths gradually brought [up] more dense materials, this would explain t[he] mass concentrations, or mascons, in the maria [as] well. The great depths to the offset core on t[he] farside of the moon would explain why there a[re] so few, and smaller, maria there.

Thus the story of the evolution of the mo[on] is slowly being worked out, as scientists analy[ze] and correlate the data that have been collect[ed]. The story will be developed and revised, wi[th] many changes, for a long time to come.

19 LUNAR PHOTOGRAPHS
FROM SPACE VEHICLES

Until 1959, all photographs of the moon were made from the earth. Since then, lunar observations have accelerated at an almost unbelievable rate. The farside of the moon, never viewed from the earth, has been photographed more thoroughly than many areas of the earth itself.

The first photographs of the moon's farside were made by the Soviet space vehicle, Luna 3, which passed around the moon in an elongated orbit in October 1959. Many of the larger features of the farside could be mapped and named on the basis of the photographs of Luna 3. The farside was no longer "luna incognitum," although actually viewed by only a few astronauts.

Then American Ranger space vehicles were successfully aimed at the moon, to make a series of fine photographs of its nearside as they fell toward the moon, up to the last few seconds and miles before impact on its surface. At once the resolution of the old, earth-based photographs, made with 100- and 200-inch telescopes, was bettered a thousand-fold. A number of the striking and informative Ranger photographs are to be found on pages 164 to 167. Rangers revealed the surface features of several typical areas, as well as providing remarkable closeup photographs of Alphonsus.

Surveyor soft-landing space vehicles followed the Rangers, settling gently on the lunar surface with the aid of retrorocket engines. They telemetered panoramas of landscapes around them, sampled and dug the lunar dust and soils, and returned enough information to reassure the astronauts on solid, level landing sites clear of boulders. Pages 169 to 171 carry photographs from Surveyors.

Then Lunar Orbiters were sent into near orbits of the moon, returning hundreds of brilliant photographs, both of its nearside, its almost unexplored farside, and its poles. From so close to the moon, they revealed many features never suspected before and added to knowledge of old features, as shown on pages 172 to 175.

From Orbiter photographs came maps to be used by the Apollo astronauts who followed them, as well as the detailed maps of the whole moon, nearside, farside, and poles, to be found on pages 176 to 181. These maps are in the earth, or astronautical convention, with north at the top, not at the bottom, as when viewed by telescope.

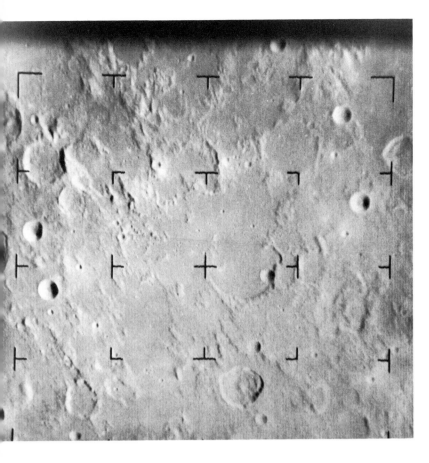

Plate 19–1. First of four plates from Range VIII. Exposed at 1965 February 20ᵈ 09ʰ 50ᵐ 45ˢ068 UT, which was 6ᵐ 51ˢ688 befor impact. Camera A. Altitude at exposure 753.79 km. Scale, 72.01 km. (Consul Chapter 15, pages 141–143.) This plate ex hibits a marked appearance of diagonal line from the lower right toward the upper lef Some able students have interpreted ther as "gashes," cut by fragments of an asteroi assumed to have impacted on the area no known as Mare Imbrium. One of them, i the lower left portion of this plate, is di cussed in some detail in the next plate. Ne the center is Lade, a large hexagon-shape walled plain, which has its two souther walls almost missing. Examination of Pla 3–11 shows that the missing walls lie with an alley of sunken terrain which exten from the vicinity of Hipparchus to pass ju south of Julius Caesar.

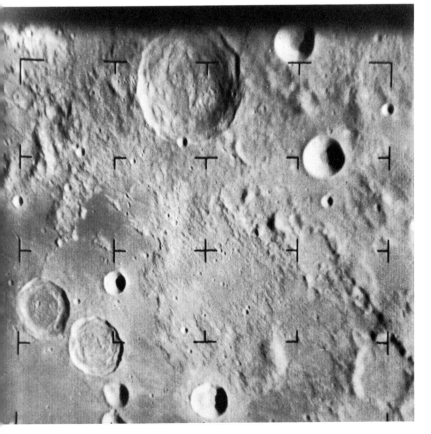

Plate 19–2. Exposed at 3ᵐ 47ˢ368 befo impact. Camera A. Altitude at exposu 412.67 km. Scale, 41.45 km. Delambre the large explosion type of crater seen at t top of this plate. D'Arrest, at the extrer lower right corner, is near the lower l corner of the preceding picture. In that p ture it was seen as connected to Delamt by one of the "gashes." Before the Rang pictures were exposed, no picture had open enough scale to examine the details this connecting line. Now careful study its parts, with attention to their shadows, veals that it is a series of ridges, rather th of valleys. Near the lower left corner i pair of large craters: Sabine with its tw Ritter, close on its lower right. Their flo are worth careful study.

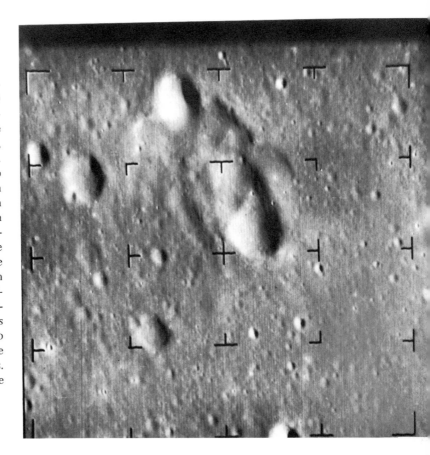

te 19–3. Exposed at 25ˢ129 before imct. Camera B. Altitude at exposure, 44.59
. Scale, 2.52 km. The length of this com-
x, peculiar feature is about 5 km. The
thern end, quite definitely, is a craterlet,
h a diameter of more than a kilometer.
e lower end, less certainly, appears also
be one. However, it seems to have an
s more than 3 km lengthwise on the main
ture. More probably this end also is an
linary craterlet with another feature im-
sed on its northern wall. Outside the
in features on both right and left are
lleys approximately parallel to its main
is. There is an appearance, certainly il-
sory, of a veil lying over the central por-
n, leaving most of the two main craterlets
covered. This "veil" reduced the albedo
that of the general background of the
ite, without obscuring the smaller details.
obably the object is due merely to the
ance association of unrelated features.

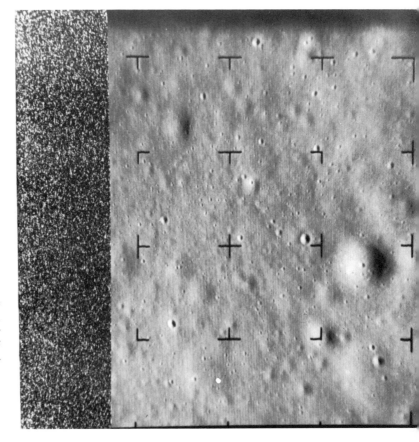

late 19–4. Exposed at 2ˢ089 before impact.
amera A. Altitude at exposure, 3.69 km.
cale, 0.42 km. Impact occurred when the
anning was about three-fourths completed.
he larger, shallow depressions, or craters,
o not appear to be results of impact but
ther to have resulted during the solidifica-
on of a once-liquid surface. There are,
owever, many sharp, tiny craterlets for
hich we do not have sufficient data to
hoose between the two hypotheses of ori-
in. These tiny pores do not appear to be
orrelated to the first class. They may ap-
ear inside or between them. We can be
onfident that these smallest craterlets are a
ater formation than the larger type and
ame into existence after hardening of the
urface. A few of them have bright halos,
robably miniature ray systems. Possibly
hese are the youngest, but a hypothesis of
mpact by a meteorite of highest velocity
ppears to be as reasonable.

Plate 19–5. Ranger IX impact, 1965 Ma 24ᵈ 14ʰ 08ᵐ 19ˢ994. Exposed at 18ᵐ 36 before impact. Camera B. Altitude, 235 km. Scale, 97.84 km. This is the first Ra IX photograph made with the 76-mm-fo length TV camera. The altitude is not m less than the maximum for Ranger pictu Alphonsus, within which lies the im point, occupies the upper right corner. eral relationships to the larger Ptolema are easily observable. Each is, appr mately, a hexagon. They have one side common. The floor of Alphonsus is divi in half by a ridge. The line of this ridg continued northward, becomes one of walls of Ptolemaeus. A long valley, one the most conspicuous of the so-ca "gashes," runs southward from the lo right corner to merge with the inner par the right wall of Alphonsus. It is import to notice that it is parallel to the Alphor ridge and to the right wall of Ptolema A line of craterlets, continuing the northern wall of Ptolemaeus, merges w the corresponding wall of Albategn None of these relationships completely c tradicts an impact hypothesis of their gin, but they are data that are hard to plain under it.

Plate 19–6. Ranger IX, camera A. Imp of Ranger IX, 1965 March 24ᵈ 14ʰ (19ˢ994. Exposed at 1ᵐ 35ˢ13 before impa Altitude of exposure, 226.61 km. Sc 23.05 km. (Consult Plates 3–19 and 16– The crater Alphonsus fills nearly the wh of the picture. Unlike such craters as Tyc and Copernicus that brighten very mu under a high sun, this one disappears exc for the famous black spots along its prin pal rill (Chapter 12 and Plate 6–1). Pr ably Alphonsus is a vast caldera, form simultaneously with Mare Nubium, whi borders it on the right. The northern h of its central ridge is seen to be very lo The bright central mountain differs in a pearance from all others on the lunar su face. The left half of the floor exhibits so very prominent rills, usually lined with c terlets. The right half shows none (see pa 138).

ate 19–7. Exposed at 1ᵐ 42ˢ812 before
impact. Camera B. Altitude at exposure,
4.59 km. Scale, 10.40 km. Plate 19–6
owed several important rills on the left
or. The greatest of them is shown in de-
il on the right side of this plate. Even a
ance demonstrates it to be so fully lined
ith craterlets that it must lie along a fault
which the craterlets are intimately re-
ted. All the rills on the Alphonsine floor
hibit this relationship more or less strongly.
ne phenomenon cannot be the result of
pacts. The cause must lie within the lu-
ır surface, at least partially. The left-hand
le of this plate is composed of the moun-
in system which lies outside Alphonsus.
hen the crests of the ridges are examined
refully, it is common to find them lined
ith craterlets. Perhaps the most conspicu-
ıs example is the one lying near the upper
ft corner. The relationship is too strong to
e the result of chance. It is impossible to
plain it by any meteoritic hypothesis. On
ne other hand, it must be remembered that
ese mountains may be considered as the
itermost parts of the Alphonsus wall. Per-
aps if we had equally open-scale pictures
f mountain systems far from any large cra-
ers we might not find such craterlets.

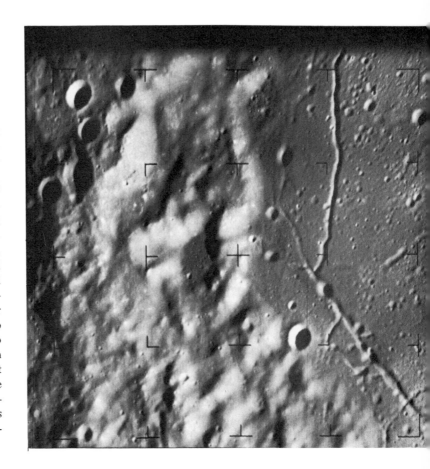

late 19–8. Exposed at 5ˢ532 before im-
ıct. Camera B. Altitude at exposure, 13.39
n. Scale, 0.57 km. This is the last complete
ame which was exposed by the Ranger IX
nger-focus camera. Impact occurred very
on after exposure of its next picture. The
idth of the whole scene is only little more
ıan 2½ km. The tiniest recognizable of the
undreds of craterlets are merely a few me-
ers across. The largest of the depressions
ıas a diameter of about 1½ km. It covers
ıuch of the area to the right of the center,
nd is so irregular in shape that it would be
nproper to speak of it as a crater. The larg-
st of the true craters lies almost at the mid-
lle of the left edge, and has a diameter of
pproximately a half kilometer. Inside of it
s a line of three conspicuous craterlets.
Craterlets with diameters near fifty meters
re very common, and a fairly definite line
f them crosses the whole picture, just be-
ow its upper edge. In general, the picture
uggests the solidified surface of a former
poiling viscous fluid. It is probable that
ome of the smaller craters were formed by
neteoritic impact, although the data are too
ew for any certainty.

Plate 19–9. Surveyor VII target area just north of the young Tycho formation, on medium resolution photograph made by Lunar Orbiter V from an altitude of 220 km. Tycho A, U, F, and T are named craters close to the rim of the main crater. The landing of Surveyor VII was planned in the hummocky area covered by the ejecta blanket thrown out with the creation of Tycho, with the idea that this might reveal deep-lying material blasted out of Tycho.

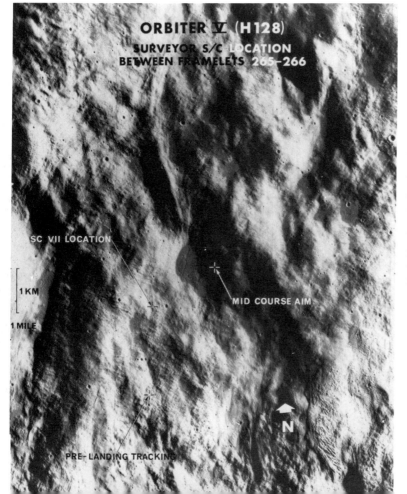

Plate 19–10. Surveyor VII landing location in hilly, ejecta-blanket terrain north of Tycho, on a high resolution photograph made by Lunar Orbiter V from an altitude of 212 km. The actual Surveyor (SC) VII location is shown between the mid-course aiming estimate and the pre-landing tracking estimate. The vehicle landed among hills of material ejected when Tycho was formed. Dark lava outflows can be seen in a number of low areas among the hills, solidifying with fissures like those seen in the floors of many other features like Alphonsus. Plate 19–16 shows the view from Surveyor VII of the lunar surface around this landing site.

Plate 19–11. This is one of the earlier pictures exposed by Luna IX. The extremely rough appearance is due to the fact that the sun was very close to the horizon and the shadows were very long. To the left of center, on the bottom edge of the picture, one of the "petal" tips is visible. Above it is a stone of perhaps 15 cm diameter. Its shadow length is at least several times its height. Still higher is another stone which actually may have a greater diameter. Beyond it is a crater which extends more than halfway across the picture. On the horizon above it is another stone. The picture shows numerous black lines, possibly similar, on a tiny scale, to the rills which can be observed from earth (pages 137–138). In the foreground are many tiny features which appear on this picture to be tiny craterlets, down to diameters of less than a centimeter. However, some of them may be merely shadows of small stones. Due to the fact that chemical, instead of solar, batteries were used, pictures were received after February 4. The height of the camera above the surface was estimated as near 60 cm. It was rotated about an axis that was tipped somewhat toward the horizon, in order to secure a panoramic series of pictures. The mass of Luna IX which impacted was about 100 kg. Under the astrographic convention the lunar coordinates of the impact point were 7° north and 64° west.

Plate 19–12. Part of Surveyor I landing pad and nearby lunar surface. Each of the three legs on which Surveyor I landed had a shock absorber on which the initial impact force, the slight rebound, and the final weight pressure were recorded. From their records and from examination of photographs, such as this one, it was learned that the surface at the impact point was a slightly cohesive agglomeration of particles usually less than a millimeter in size but larger than fine dust particles. It is estimated that the footpads sank between four and eight centimeters below the surface, rebounded slightly, then settled into final position. The surface gave evidence of a slight compression and showed that some material was pushed out from the impact depression. This picture was exposed on June 4, using the 100-mm camera focus when the solar altitude was 54°, sufficient to prevent shadows from gentle slopes. Surveyor carried no general experimental apparatus, merely photographic, plus those sensors which were needed to secure data about itself. However, especially with regard to lunar surface temperatures, these did furnish some additional data. The camera was equipped with a variable-focus lens with extremes of 25 mm and 100 mm, usually employed at one or another of these limiting values. This picture is a "narrow-angle" frame with a width of 6°.4.

Plate 19–13. Lunar surface features revealed in mosaic of photographs transmitted to earth from Surveyor III in 1967. View is due north to the horizon. Landscape is rock strewn, as in many lunar regions. Rocky, 14 meter crater appears just below the horizon with pair of large boulders in upper center. Small crater in foreground is about 1 meter in diameter. Surveyor III functioned very successfully, returning some 6300 photographs to earth, including pictures of earth during a lunar eclipse when the moon's surface temperature dropped to –155° F.

Plate 19–14. Lunar rock as it appeared in two frames of Surveyor I's telephoto TV camera lens, given special computer processing here on earth to bring out the fine details of the rock's structure. Nosing up through fine-textured lunar soil, the rock was about 5 meters from the camera and 0.5 meter long. This and many others of the 11,150 photographs transmitted by Surveyor I during June and July 1966, gave the first direct survey and analysis of the lunar surface from the landing site in the Oceanus Procellarum near the lunar equator.

Plate 19–15. Mosaic of lunar surface directly in front of Surveyor VII spacecraft that landed north of Tycho (see Plates 19–9 and 19–10). The alpha-scattering instrument on the left was designed to measure the chemical composition of the lunar surface. The surface sampler device had a scoop on the end of a motor-driven pantograph enabling it to range over an area of 24 square feet. The sampler pressed, dug into, and trenched the surface soil to a depth of 18 cm. It showed that the surface material is similar to damp, fine-grained soil or wet-hard-packed sand on earth, although of course there is no water on the moon.

Plate 19–16. Lunar landscape from Surveyor VII spacecraft in hills formed from debris ejected from Tycho. (See Plates 19–9 and 19–10.) Center of horizon is about 12 km distant. Undulating surface is characteristic of flanks of rims of many large lunar craters, with series of hills stretching to the horizon, just below which are small ravines and gullies probably formed during the original flow of debris. Rocky crater in foreground is about 2 meters in diameter and 5 meters from Surveyor's camera, with rock on near rim about 0.5 meter across. Sixty-meter crater in distance on left is about 640 meters distant. This crater has two smaller craters within it, 6 to 7 meters in diameter. Rock visible at right center may be some 6 meters across.

Plate 19–17. "The man in the Earth." The following is NASA's press release concerning this picture: "The world's first view of the Earth taken by a spacecraft from the vicinity of the Moon. The photo was transmitted to Earth by the United States Lunar Orbiter I and received at the NASA tracking station at Robledo de Chavela near Madrid, Spain. This crescent of the Earth was photographed August 23 at 16:35 GMT when the spacecraft was on its 16th orbit and just about to pass behind the Moon. This is the view the astronauts will have when they come around the backside of the Moon and face the Earth. The Earth is shown on the left of the photo with the U. S. east coast in the upper left, southern Europe toward the dark or night side of Earth, and Antarctica at the bottom of Earth crescent. The surface of the Moon is shown on the right side of the photo." Three simple facts must be remembered when looking at such a picture. The phase of Earth as seen from Moon is exactly the reverse of Moon's phase as seen from Earth. The appearance of Earth depends not only on the positions of continents and oceans but also on the seasons and the size and severity of its storms. The surface of Earth always will be more or less blurred by the effects of even a clear atmosphere.

Plate 19–18. On first examination of this photograph it appears to be a view of an ordinary region of the moon's mountainous areas. However, upon realization of the scale, we see that the implications are different. The width of the picture approximates 1350 km and the length 1500. The circular crater near the southern edge has a diameter of nearly 400 km. This is larger than that of any similar feature on nearside. Even Clavius and Deslandres are dwarfed by it. It is larger even than some of the smaller maria of nearside, for example Mare Humboldtianum. Just to the north of this giant, and slightly to the left, is a curved scarp with a rather flattish area at its base. The principal part would give the impression of being part of an immense ringed plain. On its left the ring is completely interrupted by a line of five moderate-sized craters. To their left is a quite large, typical ringed plain. Between the scarp and the 400-km crater is a small, deep, ringed plain with a central mountain. Whatever may be the cause of ringed plains, it must be a process which sometimes can be interrupted before completion. (See Chapter 12.) We can guess that this area is primarily the result of endoforces, not of impacts, although probably there were many small ones.

Plate 19–19. Tsiolkovsky, a great farside feature named for the pioneer Russian scientist who worked out the principles of rocket propulsion and space travel. In this view by Lunar Orbiter III, from an altitude of 1463 km, much of the floor is seen to be covered with dark mare material, which must be young because it has so few large craters. Compare this fresh crater with Tycho, Plate 19–9. Both show terraced walls, a central peak, and rough, creviced floors, but the dark lava flow did not enter Tycho. Another bright feature, Aristarchus, reveals the same kind of floor in Plate 16–14. Scale, 7 km, from one framelet line to the next.

Plate 19–20. One of the few maria on the farside, Mare Moscoviense was named by Soviet scientists in honor of the city of Moscow when they discovered it in the first farside photographs from Luna 3. This view by Lunar Orbiter V, from an altitude of 1236 km, reveals many ghost craters on the mare, which is about 350 km in diameter. On the left, the dark mare material nearly fills the depression, covering most detail. On the right, much of the detail remains where the depression is incompletely filled. Scale, 6 km, from one framelet line to the next.

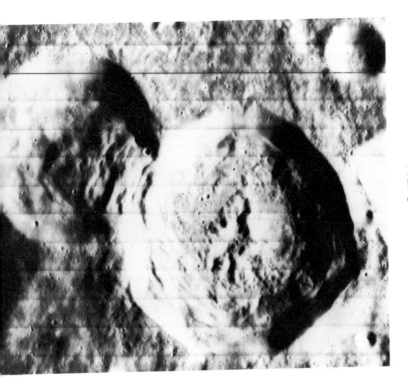

Plate 19–21. A larger and later-formed crater breaks into a smaller one in this farside Lunar Orbiter picture. (Compare with Plate 16–7.)

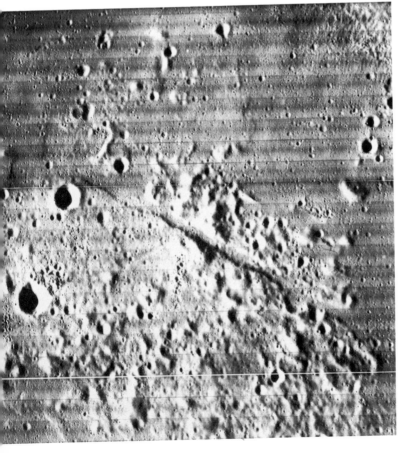

Plate 19–22. This valley viewed by Lunar Orbiter resembles the Alpine Valley so closely, with the exception of the central rill, that they must have formed in the same manner. (Consult Plate 15–6 and page 141.)

Plate 19–23. This plate from farside must rank among the most important photographs of lunar features. The area is located at 146° east (astronautical convention) longitude, well around toward the point which commonly is farthest from what is visible on earth. The picture approximates to 80 km in width and almost 95 km, north and south. The larger of the two principal craters has a diameter of about 40 km, the smaller of 25. Superficially, at least, the pair resemble Alphonsus and Arzachel. In this case there is no need to avoid a definite statement that both have, in the main, resulted from endoforces and that each is primarily a caldera. (Plate 16–1.) The larger exhibits a long rill which is nearly parallel to the inner base of its rim. Like the corresponding rill in Alphonsus, it contains small craterlets. A low central ridge crosses the floor as in the case of Alphonsus. At the extreme north and left of the floor there is a short, heavy rill, similar to those on the floor of the smaller southern companion. The floor of the southern crater shows more rills and wider ones than, possibly, any other lunar crater. We can be certain that at one time it was completely molten. After this it solidified and was acted on very strongly by one or the other of two forms of endoforce. Possibly the still-molten interior formed craters from below as in the case on the island of Mauna Loa, which exhibits also the reverse case where the lava has been drained away, causing cracking and collapse of the solidified surface.

Plate 19–24. This farside plate, about 1080 km wide and 1280 km in north-south height, shows Mare Orientale at top (Plate 12–3). The first object to catch the eye is the immense walled plain which carries also some of the characteristics of a ringed plain. (See crater plains in glossary.) Its diameter is roughly 330 km. The sun was 20° above the horizon at the center of the plate. It was exposed on August 25. Plate 19–24 was exposed earlier on the same day. Perhaps the object worthy of the most study is located just below the middle of the southern edge of the picture. It is quite definitely a ringed plain, the central part of which was molten at some past time. Halfway between it and the right-hand edge is a scarp which is centered on this ring. Apparently there was some general sinking which was insufficient to melt the surface except at its center where a strong local subsidence created this plain. It is smaller than Mare Nectaris and the related Altai Scarp but the resemblance is sufficient to indicate the same sort of origin. (Consult page 93.) The diameter of Mare Orientale is approximately 230 km.

LUNAR EARTHSIDE CHART

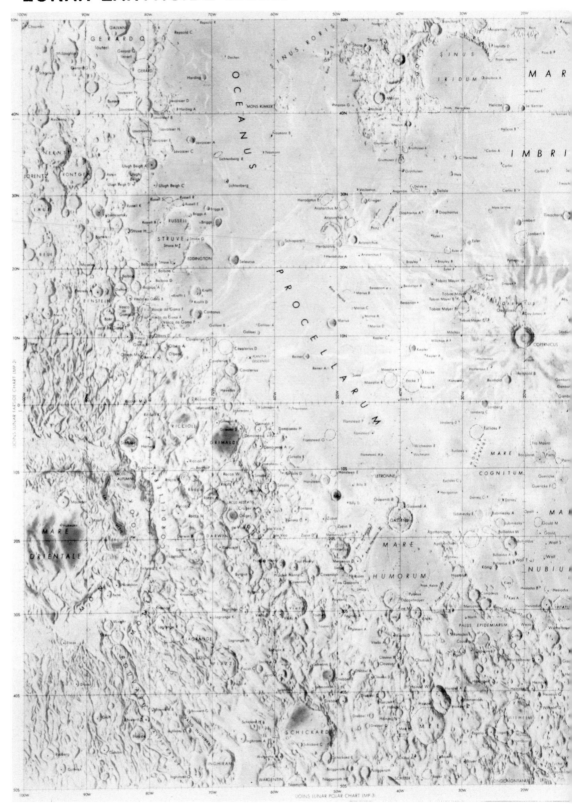

Plate 19–25. Lunar Earthside Chart based on Lunar Orbiter photographs. Made in the common earth or astronautical convention. North is at the top and west at the left, the opposite of map in Plate 3–1.

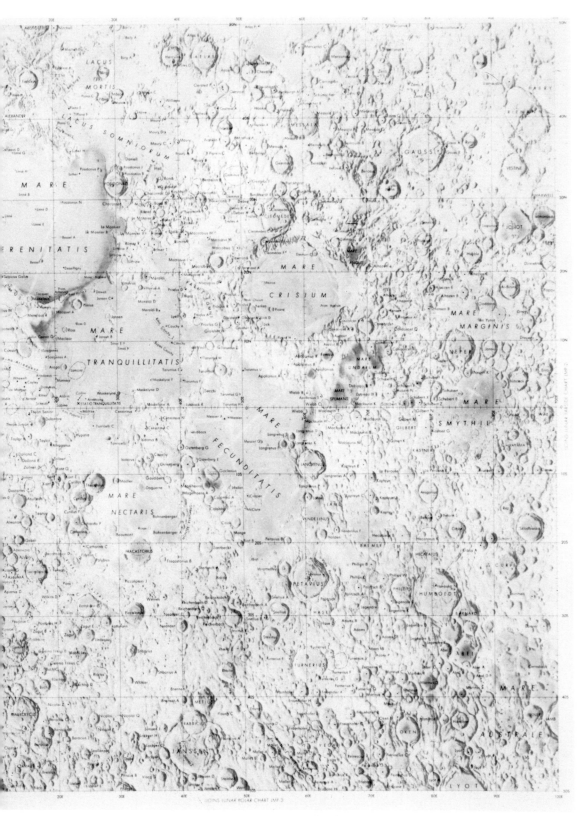

Plate 19–26. Second half of Lunar Earthside Chart, made by the National Aeronautics and Space Administration. The scale on each of lunar maps, Plates 19–25 to 19–30 is $10° = 60$ miles or about 94 km.

LUNAR FARSIDE CHART

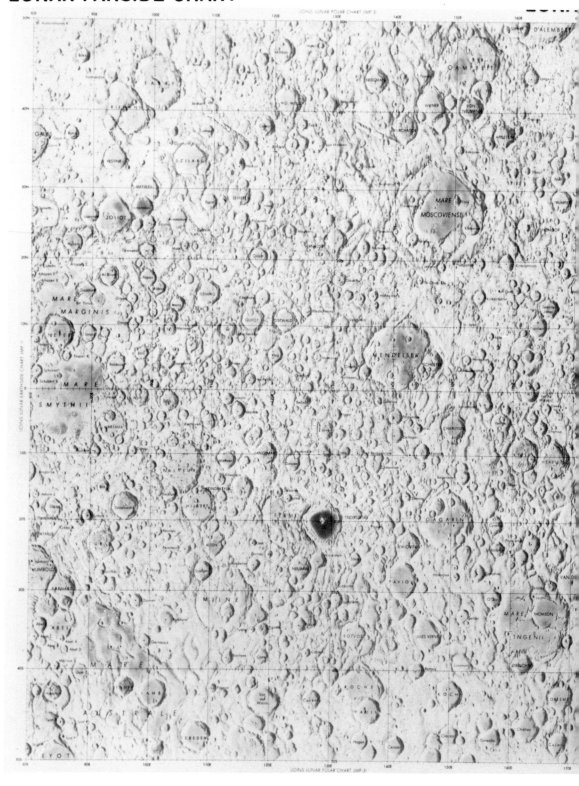

Plate 19–27. Lunar Farside Chart, based on Lunar Orbiter photographs, depicts many features never directly viewed by man, although Apollo astronauts saw some of them in their near-equatorial orbits.

Plate 19–28. Second half of Lunar Farside Chart. Many of the farside features in Plates 19–17 through 19–24 can be picked out on these farside charts. Scale is 10° = 60 miles or about 94 km.

LUNAR POLAR CHART

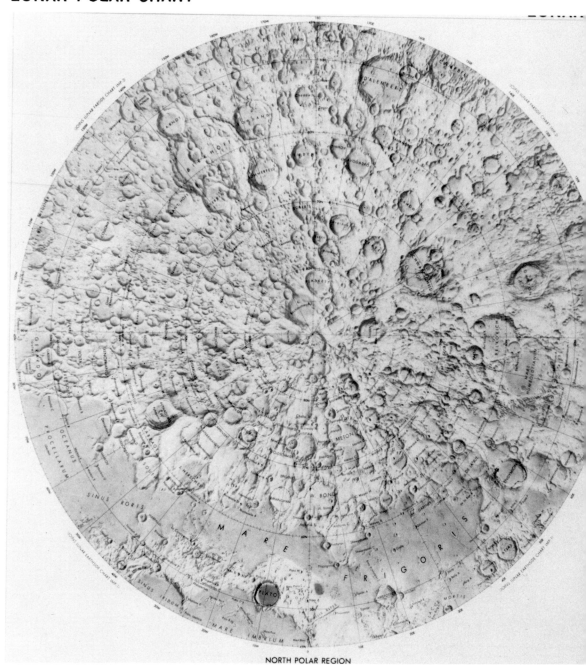

NORTH POLAR REGION

Plate 19–29. North polar region of the moon, based on Lunar Orbiter photographs, covering both near and far sides. Note craters Peary and Byrd, near the North Pole, although many features are not yet named.

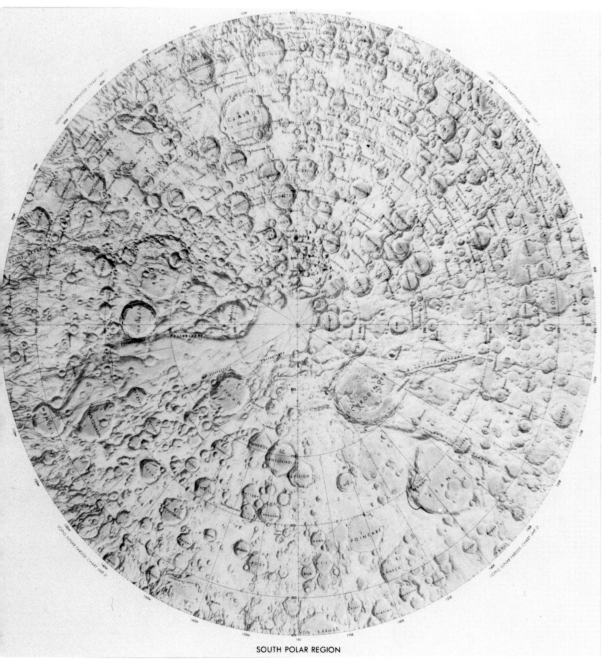

SOUTH POLAR REGION

Plate 19–30. South polar region of the moon. The blank area around the pole, within the dashed line, is not actually a lunar plain, but an area for which the photography was unsatisfactory.

20 LUNAR PHOTOGRAPHS
FROM THE SURFACE

The Apollo astronauts opened up an entirely new world, the territory of the moon, and took us all with them by TV and the hundreds of photographs they made of the lunar surface.

First, after breathtaking descents in their lunar modules, they saw strange new landscapes all around them, like the panoramas shown on Plates 20–2 to 20–5. Gently rolling hills and ridges rose to smooth mountains in the distance. They stepped down into inches of fine gray lunar dust, churned through ages of impacting meteoroids. The dark, shiny glass particles in the dust at first deceived them. It looked wet. But the moon is a waterless desert never approached on earth. Getting used to one-sixth gravity, they stumbled over rocks and craters, from small to large, all around them. Shadows under the blazing sun were inky black in the vacuum. And above the horizon the sky was black. All the sharp contrasts emphasized the stark extremes of the lunar surface.

Then, beginning to explore the moon, the astronauts looked into everything around them. They sampled dust and raked up pebbles. They hammered off rock specimens and drove core tubes deep below the surface. They set up instruments and rode around in their Lunar Rover to visit many different sites in the area. And each Apollo mission yielded hundreds of photographs.

On the Apollo 15 and 17 missions, the landings were made in the midst of mountains, over which the astronauts flew in. Fortunately, the lunar module proved capable of precise maneuvering and pinpoint landings. Plate 20–1 shows a view from Lunar Orbiter V of the mountainous region in which Apollo 15 landed. The mission investigated the mountains, the nearby Hadley Rill and the cratered floor of Palus Putredinis around the landing site. Plate 20–6 gives the first view ever down a rill—Hadley Rill. Plate 20– shows a telephoto closeup of Mt. Hadley. The observations and photographs of the rill, the mountain, and the mare increased our knowledge of such lunar features by leaps and bounds.

For years to come, scientists around the earth will be fitting together all the data from the Apollo missions to explain the rills, the maria, the mountains, and the craters, all the strange features of the moon. The photographs, thousands of them, will be studied painstakingly, to add their bit to our expanding knowledge of our only natural satellite. More is being learned at the same time about the satellites of other planets and the planets themselves. The broader theories and principles furnished by this knowledge will also be productive in the development of the understanding of our own planet earth.

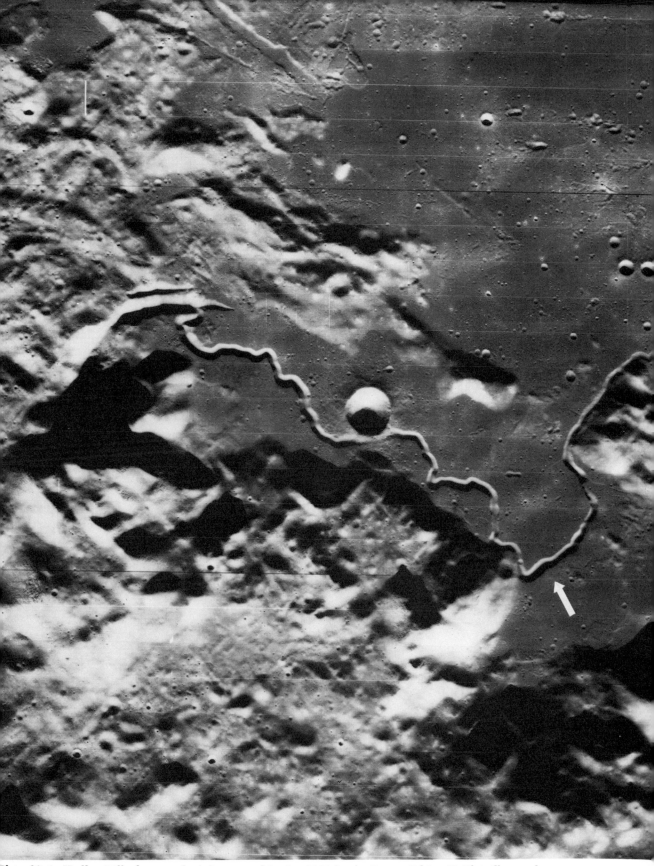

Plate 20–1. Hadley Rill photographed from an altitude of 131 km by Lunar Orbiter V. The rill meanders past Hadley C crater between Palus Putredinus (left) and the Apennine Mountains (right). Apollo 15 astronauts landed (arrow) north of the angle where the rill turns northwest, between the Apennine Front and Mt. Hadley. The Apennines rise 1 to 2 km above the mare. The rill measures from 1 to 1.5 km in width from rim to rim. It was one of the main objectives of Apollo 15 exploration, since the formation of such rills has been a great mystery.

Plates 20–2 (above) and 20–3 (top of page 185) show panorama of lunar surface taken by Apollo 15 astronauts in Apennine Mountain region (see also Plate 20–1). Mount Hadley is the high peak at the right (Plate 20–2). Astronaut in center is reaching to pick up the lunar surface drill for a core of top layers of surface and to measure surface heat flow in hole. On left is central station of Apollo Lunar Surface Experiment Package (ALSEP) furnishing nuclear power for six experiments to which cables run. Instruments include seismometer to record moonquakes and meteoroid impacts (Figure 20); magnetometer to measure the lunar surface magnetic fields; solar wind spectrometer to measure the solar protons and electrons reaching the moon; ion detector to check the density of ions in the solar wind; ion gauge to measure any very thin lunar atmosphere; and heat flow experiment to measure heat flowing from moon's depths to the surface.

Lunar Rover is parked in front of Lunar Module in left-center of Plate 20–3, with unnamed Apennines beyond. On right is Mount Hadley Delta, with St. George crater dark near its base. Hadley Rill passes to right of St. George crater, out of sight in this plate. On their first expedition in the Lunar Rover, astronauts look straight up Hadley Rill from the base of St. George crater and made the photograph appearing in Plate 20–6.

ates 20–4 (bottom of page 184) and 20–5 (below) provide a panorama of lunar landscape taken by Apollo 16 astronauts in the lunar highlands on a rocky, undulating plateau north of the ancient crater Descartes. The landing site is one of the highest areas on the moon on the side facing the earth, a good 2000 meters above the Mare Tranquillitatis on which Apollo 11 astronauts had made the first manned landing.

The Lunar Rover is parked in the center of the plateau (Plate 20–4) while the astronauts, one of whom stands behind the Rover, explore the area in the vicinity of North Ray crater on one of their traverses. The large, lone boulder on the right was named Shadow Rock. Smokey Mountain rises in the distance on the right.

Plate 20–5 (below) continues the spare plateau view of Plate 20–4. The Descartes Mountains can be seen far off on the lunar horizon at left center. A group of large boulders partly buried in the lunar soil is seen in the center of the photograph. It was suspected that the plateau viewed here, called the Cayley Plains, might be the result of an early large volcanic flow, but this did not prove to be the case when the soil and rocks from this area were analyzed. Although the plateau materials were very old, they were not specifically of volcanic origin.

Plate 20–6. Apollo 15 astronaut beside Lunar Rover
looks down the winding Hadley Rill, photographed
by the other astronaut on the rise to Elbow Crater.
Rock-strewn slopes of rill show signs of horizontal
layering which indicates complex lunar crust
formation.

Plate 20–7. First closeup view down the side of
rill, taken of Hadley Rill by astronauts of Apollo
mission. Many definite horizontal layers, or stra
appear along the side of the rill, indicating that t
lunar crust was laid down here in many flows
depositions.

Before astronauts landed on the moon, a number of theories for the lunar rills had been proposed. Some thought rills to be graben, or deep gullies, formed by the dropping of blocks of crust along faults or lines of weakness in the crust; others thought them formed by upthrusting dykes creating the rills between them. Another theory was that the rills, particularly those of a sinuous variety, had been created by fluid erosion, of water, molten lava, or gas, with gradually eroded channels. Another theory suggested that rills were formed when empty lava tubes collapsed. Rills also were thought to be collapsed crust along chains of craters; more craters can be seen along many rills than would occur by chance.

Although no water has, as yet, been found on the moon, so the running water theory of sinuous rills has had to be discarded, most of the other theories are still alive. Some of the rills may well be graben, which look very like those found on the earth. Others may be collapsed lava channels, or collapsed chains of craters formed along lines of structural weakness in the crust. Or perhaps a new theory will be formulated to explain the lunar rills as more specific knowledge of lunar processes is gained.

The discovery of layered structures along the Hadley Rill (Plate 20–7) and on Hadley Mountain (Plate 20–8) solidly confirmed and enlarged what had been suspected from cores that had been drilled from the lunar crust. These crustal areas, at any rate, have been built up gradually at some time in the past, layer on layer, from flows or from compaction of crustal materials. This stratification complicates the geologic history of the moon.

Plate 20–8. Telephoto view of Mount Hadley made by Apollo mission astronauts. The blocky structure and the horizontal linear organization which stand out clearly here indicate that the Apennine Mountains must be composed of many strata or layers of a very complex character.

Plate 20–9. Taurus-Littrow area of moon as surveyed by astronauts on their third and last traverse of the Apollo 17 mission. Taurus Mountains and Littrow craters surround an eastern bay of Mare Serenitatis. Mountains on horizon rise about 2 km above the darker plain, on which craters are scattered. Some mountain slopes exceed 25°. Size of split lunar boulder can be judged by 6-foot astronaut. Samples from boulder will tell its composition and age. Was debris on lower apron of rock eroded from its higher surfaces or deposited with the blast from meteoroid impacting nearby?

The scene in Plate 20–9 encapsulates manned exploration of the moon to date. The scientist-astronaut, Harrison Schmitt, is setting down a gnomen to indicate the angle of the sun, the scale, and the lunar color for a photograph of the split boulder. The light-colored lunar highlands, sampled in an earlier traverse, will reveal the materials and the age of these most ancient formations. Note the horizontal bands that can be dimly picked out, one above another, on the mountain on the left. Was this mountain not formed from many strata?

The plain is mantled with darker materials that may well be of volcanic origin in this Taurus-Littrow area. They may shed light on the thermal history of the moon, with volcanic eruptions or residual outgassing. Chunks and photos of the boulder may reveal more about the solid rock normally underlying the lunar soil. This is the way data from the moon have been collected and brought back to earth up to the present time.

Geologic explanations of areas like this based on principles used to understand earth processes seemed to fit quite well until men actually landed on the lunar surface and began to study it. Close scrutiny of the surface and intensive studies of the materials brought back to earth have shown that the moon follows unique principles and processes of its own, however. Features just like those shown in Plate 20–9 have never been seen on earth and must be studied in their own right.

21 MAN'S FUTURE ON THE MOON

The reasons that impelled nations and their peoples to develop space programs that eventually landed men on the moon were many. Perhaps the most compelling considerations were those of power politics and international prestige, but these need not concern us here. The fact was that the moon was there and the means to reach it were within men's grasp. Unless the drive to explore and discover and understand is forcibly restrained, men will go back to the moon eventually to continue the exploration of our satellite that has only just begun.

Eventually the time will come to settle down in semi-permanent stations. One likely site for such a settlement on the rocky lunar surface would be a small crater-like depression. Such a craterlet if properly roofed would offer maximum protection against the temperature changes of the month and would be relatively invulnerable to meteoritic bombardment. It would also provide a partial necessary radiation shield. It should be rather easy eventually to roof over such craterlets in a way that would be almost impossible of accomplishment on earth. The small weight of a domed roof should make easy installation possible; the air pressure inside the living quarters should support the whole weight. Above the roof dome there would have to be a

second thin roof to protect the interior from solar radiation, which would otherwise at times beat upon the quarters with an intensity beyond human endurance.

Within the first small craterlet "village," man can relax and live somewhat as he does on his home planet. If this first experiment is successful, larger craterlets will be prepared for village sites. Lines of craterlets such as exist in the vicinity of Copernicus may be useful. One would guess that because of safety conditions and engineering difficulties it is unlikely that any large crater will be prepared for habitation. "Cities" very probably will be entirely unknown.

During the first few weeks of living on the moon it will be necessary to use supplies brought from the earth. In order to reduce to a minimum the mass of supplies that must be transported, air and water will be recycled and purified, and vegetation grown on the rocket itself may supply much of the food. If it should prove to be impossible eventually for man to obtain his sustenance from the moon itself, little could be done on its surface other than to make the scantiest and quickest of surveys. Indications are, however, that the moon can be made self-sustaining for a race which has great scientific and technical knowledge.

Plate 21-1. The Taurus Mountains loom behind the Apollo 17 astronaut in this photograph from the last Apollo expedition. The high-gain antenna of the Lunar Rover is aimed toward the earth, seen at half-phase in the black lunar sky. The space-suited astronaut beside the "moon buggy" typifies the high-mark of the Apollo missions, in which eighteen astronauts landed on the lunar surface. The roving lunar vehicles of the future, setting out from permanent installations under the moon's surface, will carry men within an enclosed "earth" environment across hundreds of miles of lunar territory, seeking its outstanding features.

The peak of lunar exploration to date is pictured in Plates 21-1, 21-2, and 21-3, made in 1972 during the last Apollo expedition. The astronaut in Plate 21-1 was in immediate communication not only with his fellow astronaut, who made this photograph, but also, through the antenna above him, with the Manned Spacecraft Center in Houston. These expeditions on the moon lasted from 5 to 7 hours, surely a working day.

However crude their pioneering equipment on the moon, these astronauts made many remarkable additions to man's knowledge. A moment of discovery is pictured in Plate 21-2, just as the astronauts had driven up to the brim of Shorty Crater, which slopes down on the right. The whitish material around the Rover and off to the right along the crater rim is the "orange soil" discovered there. Plate 21-3 pictures this thin layer of reddish-orange soil in closeup. They also saw reddish streaks extending down into Shorty Crater. However, the crater did not have the cinder cone in its center they had hoped to find—only a mound of rocks.

The analysis of this "orange soil" is under way. Mostly it consists of an orange-brown glass,

but not of the kind of volcanic glass created when water is present, as in earth volcanoes. Its age is 3.7 billion years, going back to the time when the lunar crust was probably very hot. Yet no full explanation of this "orange soil" has been offered. The lunar process by which it was produced in this location is still unknown.

This mystery, and many others, about the moon may be cleared up, and may not, as scientists continue to work with the data at hand. But full knowledge of the lunar surface, its interior, and its history can only come with further explorations, both unmanned and manned, of all its peculiar and distinctive features.

Someday the equipment pictures in these plates and the techniques of the astronauts will seem as out of date as the "horse and buggy" era on earth does now. Villages will be growing here and there across the moon, and colonists will spend long periods of their lives there. They will come to make it their home and find it beautiful, just as the early American pioneers felt as they carried the frontier further and further west.

A lunar area well worth much more investigation is pictured on Plate 21-4 on page 192.

Plate 21–2. In their third traverse in the Lunar Rover of Apollo 17 mission, astronauts discovered the "orange soil," which appears whitish near the Rover and further along the crater rim.

Plate 21–3. Nearby the rock outcropping to right of Lunar Rover in Plate 21–2, astronauts made this photograph of the strange "orange soil," showing whitish in a thin layer in front of the tripod.

Plate 21–4. Closeup of southern portion of Aristarchus made by Lunar Orbiter V from an altitude of only 129 km. Crater floor at upper right is rough and full of pits. Black specks are boulders that have eroded the terraced walls. Rim (lower left) also contains many holes or pits.

Part of the floor, the terraced walls, and the rim of Aristarchus are shown closeup in Plate 21–4, and a broader view of the vicinity in Plate 16–4, both made from Lunar Orbiters. The Apollo 15 command-service module carried an alpha particle (proton) instrument to locate regions on the moon with radioactive radon activity. The hottest radon spot observed was Aristarchus, which scientists believe indicates escaping gases—the residual outgassing of Chapter 16. The sighting of transient lunar phenomena in and around Aristarchus has been frequent. When man returns to the moon, Aristarchus merits investigation. The area might offer many materials not found elsewhere.

Hopefully man will use the moon solely as an international laboratory for basic scientific research. The most exciting prospect one can see at this time is in astronomy. The moon is an ideal site for an observatory. Astronomers at the surface of the earth cannot possibly use their telescopes with any great efficiency; as a matter of fact it is really surprising that in three and a half centuries they have been able to do as much as they have. The reason for this is simple enough: we live at the bottom of a great ocean of turbulent air.

This ocean of air absorbs roughly one-half of the incoming radiation from the sun and from other objects, and in particular it absorbs almost all of the very short-wave (or high-frequency) radiation. This is in itself an extreme handicap to the astronomer, but it is not the worst of the story. The worst problem is turbulence. Our air is composed of various components or masses continually moving and mixing at different temperatures. Air at one temperature and density bends light passing through it differently from air at another temperature and density. The result of this hodgepodge is that all objects in space appear to tremble; the twinkling of the stars is a simple illustration. Telescopes magnify this shimmering effect as much as they do the objects under view. Even under the best conditions at such places as Mount Hamilton, Mount Wilson, Palomar Mountain, and Kitt Peak, this effect is always very serious.

Because of these innate disadvantages it will be well to consider the potentialities of an observatory located on the surface of the moon and the advantages that may be derived from it in our effort to learn the organization of the physical universe. The list of advantages given below is extremely incomplete, for there is no point at which some benefit will not be apparent for an observatory.

The accuracy of star definition is important for many investigations including fundamental observations of the distances of stars. These are obtained by photographing the nearer stars against a background of the more distant ones at time intervals of about six months. During these half-year periods the position of the earth has been shifted by 186,000,000 miles by reason of its orbital motion around the sun. As we look at a nearer star from these two opposite sides of the orbit, it appears to have moved its position slightly with respect to the more distant stars behind it. The star's apparent shift from its mean position is called its *parallax*. For Alpha Centauri, the closest of all stars, the parallax is about three-fourths of a second of arc. Very serious errors in the measurements of these small angles, and therefore in distance calculations, result from the fact that the images vary in apparent size because of atmospheric distortion. If we could secure nearly perfect images, our measurements of parallaxes would be much more accurate, and we could carry our direct measurements of distance to a far greater number of stars. Our

whole scale of distances in the universe depends upon these fundamental measurements. On the moon, where the atmosphere is negligible and conditions always ideal, we would obtain a far more accurate fundamental scale of distances.

Because of the almost perfect images that will be formed when a telescope is used on the moon, all the light from a star will be concentrated on a smaller area of the plate and will therefore blacken it in a shorter time. Moreover, with no atmosphere to absorb stellar light, the stars will appear about twice as bright as they do on earth. Finally the long nights and the very important absence of atmospheric glow in the night sky will make possible photographic exposures many times longer than on earth. The net result of these three factors will be the observation of dwarf stars much farther from the sun than any that can be seen at present. Even if the maximum distance visible were to be multiplied by as little as ten times, a sample 1,000 times greater than is available from the earth would become accessible.

Even the weaker gravitational attraction on the moon will be an important advantage. The largest refracting telescope in the world, the 40-inch at Yerkes Observatory, just begins to exhibit a distortion of images as the result of objective bending under the earth's gravitational pull. The 200-inch mirror at Mt. Palomar has been made very thick and is supported at many points by a very complicated lever system in order to prevent bending that would distort the stellar images. The tubes of telescopes also bend under their own weight, and expensive special construction is the unfortunate but necessary result. At the surface of the moon, where objects weigh only one sixth as much as on earth, there will be far less distortions of stellar images due to flexure in the telescopes. And because telescopes there will be much lighter in weight, they will be easier to handle and cost less to build.

Study of the sun's corona, or faint outer layer, will be more effective from the moon. On earth these studies are best advanced during total solar eclipses, when the moon hides the entire bright surface of the sun from a small area of the earth. During an eclipse the far fainter outer atmosphere or corona of the sun can be observed protruding beyond the hidden disk of the solar photosphere. Under unusual conditions the co-

Plate 21–5. Solar corona as seen from Apollo 15, one minute before sunrise above the lunar horizon. From the moon, the corona and many other celestial objects could be studied most efficiently.

rona has been observed to extend more than 10,000,000 miles from the sun's disk. But even during the best of eclipses the earth's sky is rather bright, and the astronomer must observe the faint, pearly streamers of the corona against a background of such light. Naturally a great deal of detail is lost. But a very simple form of the coronograph, used principally to keep extraneous light away from the photographic plate or other sensor, will produce observations of the coronal streamers far more satisfactory than those obtainable here at any eclipse. The fine details can be followed hour after hour, for periods extending through the long lunar day. With details of coronal changes observed continuously, and with its spectrum uncontaminated by terrestrial light, the manner in which the coronal atoms radiate or scatter light can be studied much more satisfactorily. The streamers of the corona will be followed much farther from the

sun than would ever be possible on the earth and possible segregations of various atoms in different outer regions of the corona can be looked for.

Our atmosphere interferes with the passage of other than visual radiation, of course. It seems reasonable that the short-wave solar radiation absorbed by the atmosphere must influence our weather far more than do the forms of radiation to which our air is almost transparent. This short-wave radiation includes ultraviolet and probably extends to frequencies beyond those described as x-ray radiation. From the moon the solar spectrum will be observed under perfect conditions, already hinted at by the advances secured from spectrographs carried by rockets. If there is to be any hope of long-range weather prediction for the earth it seems probable that it must come in part through daily study of the sun's ultraviolet radiation.

One of the most important observations ...de by astronomers concerns the brightness of ...rs, and these too will be more effective from ... moon. Relative stellar brightnesses are meas-...d with great accuracy by the use of modern ...ctronic methods in conjunction with telescopes. ...t there is a complication. As a star rises from ... horizon toward the zenith, its light passes ...ough less and less of our air and we get a ...rious change in its brightness. Allowances ...st also be made for the varying transparency ... the atmosphere even at a constant altitude ...d during a very clear night. The changes, ...oreover, are not the same for all wavelengths ... light. The observer of the brightness of stars ...st correct all his observations for all of these ...anges, and since the corrections can never be ...plied perfectly they increase the uncertainty ... the results. On the moon very tiny variations ... the intensity of radiation from a star will be-...me rather easily observable, and photometric ...asures will have an accuracy to which terres-...al results can scarcely be compared.

On the moon the astrophysicist will be able ... penetrate obscuring galactic clouds of very ...n gas and dust more efficiently than he can ...re. Although some of these are so nearly trans-...rent that stars can be observed through them ...m the earth, especially by use of infrared ...ht, the apparent luminosities of the stars de-...ase in the process. The blackness of the lunar ...y will increase the possible length of photo-

graphic exposures made from the moon; this, in combination with the improved images of the stars previously described, will permit observation of more stars through these thin galactic clouds, to the great benefit of the astrophysicists.

Finally, from the moon we will see far deeper into remote space. If the background were truly black, the increased length of exposure should show much fainter galaxies than can be found today. The contrast between a faint galaxy and a truly black sky would also help much in reveal- ing them.

If we are to have any hope at all of really understanding the general nature of our physical universe, questions of this kind must be answered not by mathematical speculation alone but pri- marily by definite and precise observations. There is little hope of enlightenment from this side of the moon. Man must begin to live on its inhospitable surface, carrying with him the lo- calized environment to make continued existence there possible, and the curiosity and skills and tools to enable him to initiate the great work of discovery. Those who have faith in the human race feel certain that this will be done, and in- creasing numbers believe that occupation of the moon will be an achievement of this century. Now that man has landed on the moon, what he does when he settles on it will be merely a pre- lude to what he will do later. Humanity stands now on the threshold, ready to cross soon into a new age of exploration.

Plate 21–6. First view of the hydrogen corona around the earth, photographed by Apollo 16 from the moon, with ultraviolet exposure times of 5, 15, and 60 seconds, from left to right.

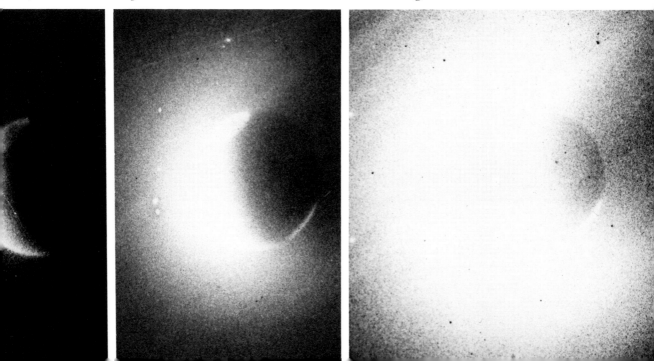

Plate 21–7. Lunar module (right) of Apollo 17 mission carrying astronauts to rendezvous with command and service modules (left) after exploration of Taurus-Littrow area. Odd noncentral peaks of crater and long valley to left of command module typify the mysterious features of the moon.

The Apollo manned-landing explorations of the moon closed with Apollo 17. The rendezvous of the returning astronauts above the moon, preparing to go back to earth, is shown in Plate 21–7. With the Apollo missions, mankind's vision was lifted from earth's horizons, viewed by men for at least 5 to 10 million years, to the acutely curved horizons of the moon, with black sky above, viewed by astronauts for three years, but captured in their photographs for all mankind to see.

As this chapter implies, men and women will be back on the moon, sooner or later, to carry their explorations further. The moon will be colonized for many purposes: To check further on theories about its origin and development, bearing on those of the earth and the whole solar system as well; to support astronomical studies of the types mentioned here; to study human reactions to low-gravity conditions, which could have many applications, not as yet imagined; and to use the ultra-high vacuum of the moon for industrial research and operations. All these seem valuable.

Finally, people will colonize the moon simply because it's there, still mysterious. Life seems to probe and stretch beyond its known limits into the unknown, to all the mountain peaks and ocean depths of earth. Now life is reaching toward the moon and other great bodies out there in space.

APPENDIX

With the exception of some of the photographs made to correct for limb foreshortening, all lunar pictures exposed before 1954 were part of the Moore-Chappell series made at the Lick Observatory during the years 1937–1947 with the 36-inch refractor. During this interval it was customary to notify Dr. Moore and Mr. Chappell whenever seeing conditions were unusually good and the moon was available. They would make several exposures, then return the instrument to the regular observer for the night. Lunar and planetary photography requires unusually steady seeing conditions. Many nights that are quite satisfactory for spectroscopic work, for example, would be wasted if given over to this type of program. A plan such as this is the only efficient one to use.

Each of the Moore-Chappell photographs covered the whole of the moon. As a result, they were used for all general atlas types of work. The refractor is handicapped, of course, by lack of complete correction for chromatic aberration. For this reason all of these Lick photographs

were limited to the yellow-green region through use of a suitable filter. Most of the exposures were on Kodak Solar Green spectroscopic plates. Exposure times, depending on phase, varied from 0.1 to 1.2 seconds.

The cooperation of Dr. C. D. Shane, former director of the observatory, is much appreciated. He made it possible to study each of the negatives in detail and furnished transparencies of all for which further study was desired. In all this Mr. Chappell assisted wholeheartedly. Not only did he make superlative transparencies but he gave valuable technical advice.

MOUNT WILSON AND PALOMAR OBSERVATORIES

All of these photographs were made by the author as guest observer through the courtesy of the director, Dr. Ira S. Bowen. Between April, 1954, and March, 1959, 478 negatives were made. All were exposed at the Cassegrain focus of the 60-inch reflector at Mt. Wilson. Because of the difficulty of removing very heavy spectrographic equipment, the plates used were not large enough to cover the whole of the moon. The program was primarily a color-index examination for any significant differences between exposures made in the blue-violet and the infrared. One pair of such plates, exposed in the most rapid possible succession, composed an observation. In the earlier exposures one of these was a IIa-0, but in the later a II-0 was used. Both were used without filter and depended on the plate cutoff near $\lambda = 4900$. The earth's atmosphere interfered much less with the infrared photographs than with the shorter-wave ones. As a result they were used for all plates here except for the few cases in which comparison of a pair was desired. Except where otherwise noted, all were developed in a rocking tray for five minutes in D 19 at a temperature near 70° F. Nearly all of the infrared plates were Kodak 1-N, exposed through a Pyrex CS 7-69, CG 2600 glass filter, which has its 10% cutoff near $\lambda = 7200$. In a few cases, slow, fine-grained and contrasty IV-N plates were used near full moon phase in order to emphasize shade differences between rather large areas of the floors of maria. The same filter was used for them. Continuation of these observations is contemplated. It would have been very difficult to have prepared these pictures for study and for enlarged reproduction without the expert assistance of Mr. Paul E. Roques of the Griffith Observatory. A photograph of the moon is necessarily underexposed at the terminator and overexposed in other areas. If a print is to exhibit any large area to the public, this difference must be corrected. But if the original negative is ever to be used for even qualitative photometric research, it must not be modified. Mr. Roques met the problem by silhouetting skillfully while making enlarged transparencies and duplicate negatives.

CORRECTION OF LIMB FORESHORTENING

Before World War II Dr. Fred E. Wright of the Carnegie Institution of Washington, assisted by his son, Dr. F. H. Wright, developed a technique for correction of foreshortening of lunar features near the limb. He had the Eastman Kodak Company coat some fairly large glass globes for use as photographic plates. Next he projected a photograph of the moon on a matte-surface, white globe of the same size as the glass globes and focused very carefully. The image on the globe reversed the foreshortening and revealed features in almost their true form. Then he carefully substituted for the white globe one of the sensitized glass globes—in effect a spherical photographic plate—and exposed it. The developed picture showed the moon almost as it actually is. The optical planning was difficult and even the exercise of great care could not prevent defects at the limb.

Two of the seventeen Wright globes were loaned to the Griffith Observatory for public exhibition. In 1954 it occurred to the author to copy these globes from points directly over features of especial interest in order that pictures of the true forms could be available to a larger audience. Mr. Roques cooperated and we were fairly successful. Then, using the 40-inch parabolic mirror removed from the center of the 200-inch telescope, we began projecting plates onto white wooden globes and copying directly. Some of these pictures are reproduced in this book. They show certain features almost as we would see them from a rocket almost directly overhead.

Later some still better projections were made at the University of Chicago by Dr. Harold C. Urey, who made them available for use here. All of these pictures emphasize how different the apparent forms of limb features are from their true ones, making incorrect interpretations very frequent.

Table 6. Lunar data

Some of the following lunar data are given in greater detail than usually will be desired, but the user should have little difficulty in choosing a proper rigor.

MASS

In tons	8.0×10^{19}
In grams	7.32×10^{25}
In terms of earth's mass	0.01226
In terms of sun's mass	0.0000000368

DIAMETER

Miles	2160.
Kilometers	3476.
In terms of earth's mean diameter	0.27227
Mean geocentric, angular	31'07"

SURFACE GRAVITATION

In terms of earth's	0.165
In feet per second per second acceleration	5.31
In cm per second per second acceleration	162.

VELOCITY OF ESCAPE AT SURFACE

In terms of earth's	0.213
Miles per second	1.48
Kilometers per second	2.38

DENSITY, MEAN

In terms of water	3.33
In terms of earth's mean	0.6043

TEMPERATURES OF SURFACE ROCKS

Sun at zenith	214° F, 101° C
Night	−250° F approx.

MAGNITUDE

Mean of full moon	−12.5

ALBEDO 0.07

AVERAGE LENGTH OF MONTHS

Synodic	29.530588^{d}
Sidereal	27.321661
Draconitic (nodical)	27.212220
Anomalistic	27.554550
Tropical	27.321582

LIBRATIONS, MAXIMUM

Geocentric in longitude	7°54′
Geocentric in latitude	6°50′

FRACTION OF SURFACE

Always visible	41.0%
Sometimes visible	18.0%

ORBIT (data are those of Professor E. W. Brown)

Distance	
maximum	252,710 miles
minimum	221,463 miles
mean	238,860 miles
Parallax, mean	57′02″.54
Inclination of orbit plane to ecliptic, mean	5°08′43″
Inclination of orbit plane to earth's equator	
maximum	28°35′
minimum	18°19′
Inclination of moon's equator to ecliptic	1°35′
Inclination of equator to orbital plane	6°44′
Eccentricity, mean	0.0549

Mean velocity in orbit	
Linear, miles per hour	2,287; km 3,680
Angular, per hour	33′

REGRESSION OF NODES

Period of	18.5995 years
Annual change	19°.341

ADVANCE OF LINE OF APSIDES

Mean period of	8.8503 years
Annual change	40°.690

THE COLONGITUDE OF THE SUN

Space travel is giving this subject an increasing pragmatic value. It is not difficult except for the fact that the conventions which have been adopted are rather "sticky." The colongitude is merely the trigonometric complement of the selenographic longitude of the sun. That longitude of the sun equals the longitude of the point on the moon's surface for which the sun is in the zenith. We must remember that, with respect to space, the earth and the moon rotate about their axes in the same direction. We must remember also that, as seen from the north pole of the moon, longitude is measured in a counterclockwise di-

Figure 22. Colongitude of the sun. The circle is the moon's equator. P is the point which we on earth see as the mean center of the moon's visible disk. It is located in Sinus Medii. Its selenographic longitude and latitude are zero.

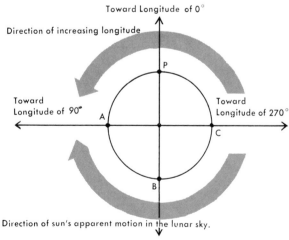

Toward Longitude of 0°

Direction of increasing longitude

Toward Longitude of 90°

Toward Longitude of 270°

Direction of sun's apparent motion in the lunar sky.

Toward Longitude of 180°

rection. As watched from this same point the sun appears to move around thé moon in a clockwise direction. This causes the selenographic longitude of the sun to change in a negative direction. The colongitude of the sun equals 90° minus the selenographic longitude of the sun. Negative values are avoided by adding 360° where necessary. In the diagram the moon is seen from a point over its north pole. The circle is the moon's equator. P is the reference point for which both the selenographic longitude and latitude equal zero. The direction of the moon's rotation is shown by an arrow. The other arrow shows apparent motion of the sun with respect to the moon.

When the sun is rising at P its longitude is 90°. Therefore its colongitude is zero, which equals the longitude of P. When it is noon at P the longitude of the sun is zero and the colongitude is 90°. When it is sunset at P the longitude of the sun is 270° and the colongitude is −180°, which changed by 360° gives the tabulated 180°.

Colongitude is useful for determining noon at a greater distance from the moon's equator than for determining sunrise and sunset. The obvious reason for this is the variation in length of time the sun is above the horizon as we go to higher latitudes.

The I.A.U. has now reversed the convention for east and west on the moon insofar as selenonautical maps are concerned. For astronomical use it has advised a use of right and left instead of east and west.

TRANSLATION OF LETTER RECEIVED FROM N. A. KOZYREV

"I am sending you prints of the three most interesting spectrograms of the Crater Alphonsus which I obtained on November 3, 1958. The slit of the spectrograph was oriented as shown in the diagram.

"Spectrogram 1) Nov. 3, 0^h to 1^h UT. You will notice that the central peak is redder than the neighboring background and has a normal appearance of the spectrum.

"While guiding there was noticed an unusual decrease in brightness. Probably at this time the central peak was illuminated by the sun and was shining through the ejected gas. [Note: on Plate

16–2 this spectrum can be seen at the bottom.

"Spectrogram 2) Nov. 3, 3^h to 3^h 30^m UT The spectrum of the central mountain shows bright gaseous emission. The most prominent emission bands are the 4756A (which has not yet been identified), and the group 4735, 4713, 4696A —the Swan bands of the molecule C_2. There are also Swan bands 5165, 5130. One sees clearly the group of bands C_3 of comet spectra between 4100 to 3950A. In addition, one can see many other bands; their identification has not yet been made [Note: on Plate 16–2 this spectrum is at the top.

"Spectrogram 3) During the guiding of the spectrogram 2 I noticed a marked decrease in the brightness of the central region and an unusual white color. All of a sudden the brightness started to decrease. At this instant the exposure was stopped and a new one was started from 3^h 30^m to 3^h 40^m. The result was a normal spectrum (there was only a slight suspicion of the Swan band 4757).

"The luminescence of the ejected gas was excited by the hard radiation of the sun which could penetrate only into the outermost layers of this cloud of gas. Hence the emission could be observed only above the little mountain, very slightly displaced (approximately one second of arc) in the direction of the sun. It is interesting that in the region of the brightest emission bands 4757–4735–4713, the shadows are deeper than in those portions of the spectrum where there are no bands. This shows that the optical thickness of the molecules of C_2 was greater than one. hence the process of the ejection of gases was of considerable magnitude and there could have been noticeable changes of the crater.

"I would be greatly interested in knowing whether you succeeded in photographing Alphonsus on December 1–2 as you intended. I should be grateful if you would send me copies of these photographs and also prints of your previous photographs of Alphonsus. If it is not too difficult, please send me also your articles on the nature of the lunar surface.

"Your remarkable photographs of Alphonsus obtained through filters, convinced me that there exist luminescent clouds of gas in this crater. As a result of this, I started systematic spectrographic observations of this crater. Your work was the first which has demonstrated the possibility of gaseous emanations on the moon."

GLOSSARY

ABSOLUTE ZERO: about 273° below zero Centigrade. At this point the normal motion of molecules, which decreases in speed with decreasing temperature, would cease completely.

ACHROMATIC REFRACTOR: see *Chromatic aberration.*

ALBEDO: the fraction of the light which a body reflects of that which falls on it.

ALTITUDE: the angular height of an object above the true horizon as measured along its vertical circle.

ANGULAR DIAMETER: the angle subtended by the actual diameter of an object as seen by an observer. Angular diameter varies inversely with distance; at twice the distance the angular diameter is half its former value. The angular diameter of the moon varies during the month from 33′30″ when closest to us to 29′21″ when farthest.

ANGULAR MOMENTUM: see *Momentum, moment of.*

APOGEE: the point in its orbit where the moon or an artificial satellite is farthest from the earth. Opposite of *perigee.*

ARGON: a heavy gaseous element that is found in small quantities in the terrestrial atmosphere (A). It is colorless and tasteless.

ASTEROID: a synonym for *minor planet:* any planetary body that is smaller than a regular planet but large enough to be observed.

ASTRONOMICAL UNIT: the mean distance between the centers of the earth and sun. It is close to 93,000,000 miles.

ATMOSPHERE: the gaseous envelope that surrounds an astronomical body. It includes the molecules of water vapor, but not the dust particles.

ATMOSPHERIC ABSORPTION: the absorption of a variable amount of the radiation that enters the air. Our air is not perfectly transparent. The absorption is greater at the blue end of the spectrum than at the red end.

ATOMIC FISSION: the breaking of an atom into lighter atoms and subatomic particles. It is accompanied by the release of intense radiation.

ATOMIC FUSION: the building of heavier atoms from those of less mass. It is the process by which the sun produces its radiant energy. In this case, on the sun, helium is formed from hydrogen.

AURORA: light emitted by the thin upper atmosphere of the earth at heights roughly between 50 and 600 miles. Auroras are caused principally by charged hydrogen particles and by electrons from the sun.

AXIS: the line through the center of mass of a body, about which it rotates. The poles of rotation are the ends of the axis, where it cuts the surface of the body.

AZIMUTH: the angular distance measured westward from the south point of the horizon to the point on the horizon directly below an object (*i.e.*, at the foot of the object's vertical circle). Engineers commonly measure azimuth either east or west from the north point instead of the south point.

BINARY: a physical double star; that is, two stars that are close together and move in ellipses around their mutual center of mass.

BLACK BODY: a body that absorbs all the radiation that falls on it and reflects none. Strictly, no object is a black body. But the moon, sun, planets, and stars approximate black bodies, and it is convenient to consider them as such in order to determine their temperatures by their general radiation.

BLINK-TYPE SCANNER (or microscope): a device in which two photographic transparencies may be adjusted side by side for examination through an eyepiece. The light that passes through them to the eyepiece is interrupted rapidly, coming from one picture for an instant, then from the other. If the photographs are exactly alike, the observer sees no change, but if some small area or object exists in only one of the pictures, or is in a changed position, or is of a changed brightness, he sees a conspicuous jumping effect instantly.

BLOWHOLE CRATER: a form of small volcanic crater found on the earth. Some of the lunar craterlets may be quite similar to them.

CALDERA: a type of volcanic crater which has been formed primarily by a sinking of its floor rather than by the ejection of lava.

CAMERA LUCIDA: a prism (or mirror) device at the eyepiece of either a microscope or a telescope, by means of which the observer sees an enlarged image on a white screen beside and below the eyepiece. This enables him to trace the enlarged picture.

CELESTIAL EQUATOR: the great circle in which the plane of the earth's equator cuts the celestial sphere. See also *Equatorial plane*.

CHROMATIC ABERRATION: the failure of a lens to bring all wavelengths of light to focus at the same distance. This produces a colored halo around the image. The fault can be partially corrected by making the lens from two or more pieces of glass of different densities. Such a lens is called *achromatic*.

CLEFT: see *Rill*.

COLONGITUDE: the lunar longitude of the sunrise terminator. Colongitude is more nearly accurate than phase for indicating the position of the sun relative to the moon because it has been corrected for the moon's librations. The *American Ephemeris* gives these values for each day.

COMET: a body of extremely low mass and density but often with a volume thousands of times that of the earth, moving in an eccentric orbit around the sun. Most of the material is in the "head," which probably is composed mainly of meteoroids and gas. Many comets have long gaseous "tails."

CONJUNCTION: the position in which one body of the solar system is most nearly in line with another, as seen from the earth. For example, the moon is in conjunction with the sun at the instant of the new moon.

CONTINENTAL SHELF: that part of the floor of the ocean near the shore of a continent and shallower than the general floor. It is bounded by a scarp which may approximately parallel the continental shoreline. Similar areas are found on the floors of some of the lunar maria.

CONTINUUM: as used here, the continuous background of light in the solar (stellar) spectrum.

CORONA: usually the outer atmosphere of the sun which, during a total eclipse, looks like a crown around the new moon. The word may be used to describe any phenomenon that looks like a ring surrounding an object.

CORONAGRAPH: an optical device by means of which it is possible to observe the corona when the sun is not totally eclipsed.

CRATER: in lunar studies, any depression of the moon's surface other than a valley. The depression may have been caused by chance or by impact of a meteorite or an asteroid; or it may be a volcanic crater or a *caldera* (sunken area). The term *endocrater* is coming into use to denote any crater that has resulted from internal causes. *Ectocrater* has been suggested for any crater thought to have been formed by external causes. However, *impact crater* covers this ground quite well.

CRATER CONE: the hill, often steep, that a volcanic crater may build about itself.

CRATER PIT: a very small craterlet.

CRATER PLAINS: the largest lunar craters are divided into two general classes: (a) the explosive *ringed plains* with their external walls and (b) the *mountain-walled plains* with little or no external walls.

CRATERLET: any lunar crater that is less than approximately five miles in diameter.

CRATERS, EXPLOSIVE: craters which appear to be results of violent explosions. They exhibit definite outer walls. These craters may be either endocraters or impact craters or possibly, in the case of certain large ones, hybrids between the two.

DAY, APPARENT SOLAR: the interval of time as measured by the sundial between successive passages of the sun across the meridian.

DAY, LUNAR SIDEREAL: the sidereal day on the moon, a time interval equal to the sidereal month.

DAY, LUNAR SOLAR: the solar day on the moon, a time interval equal to the synodic month. It varies slightly.

DAY, MEAN SOLAR: the average length of the apparent solar day throughout the year—24 hours.

DAY, SIDEREAL: the interval of time between the passage of a star across the observer's meridian and the next passage of the same star. It is approximately 3 minutes 56 seconds shorter than the mean solar day, and is the true period of rotation of the earth on its axis.

DIAPHRAGM: an opening that restricts the passage of light through an optical system.

DISPERSION: the separation of light into its component wavelengths (or colors) by a prism or a grating.

DIURNAL: daily.

DOME: a low rounded elevation on the moon. It may be as much as ten miles in diameter and perhaps as little as 50 to 200 feet high.

DRACONITIC MONTH: see *Month, nodical.*

DWARF STARS: those stars that have a luminosity comparable to that of our sun, or less.

ECCENTRICITY: the measure of the departure of an ellipse from circularity. A circle has an eccentricity of zero.

ECLIPSE: (a) the passage of one body into the shadow of another. In a lunar eclipse the full moon passes into the shadow of the earth. If it enters completely, there is a *total lunar eclipse;* if not completely, a *partial lunar eclipse.* (b) If the new moon completely hides the sun as seen from any point of the earth, there is a *total solar eclipse.* For this to occur, the moon must be at a point in its orbit where its angular diameter is greater than that of the sun. If conditions are right for a total solar eclipse, except that the moon appears smaller than the sun, there is a thin ring of the sun seen around the moon. This is an *annular eclipse.* If the new moon does not completely block the sun from view for any point on the earth but does hide part of the sun, there is a *partial eclipse.* Often total and annular solar eclipses are called *central eclipses.* The *eclipse year* is the interval of time the sun takes to move around the ecliptic from one of the moon's nodes to the same one again. The eclipse year is about 346.62 days long.

ECLIPTIC: the great circle of the celestial sphere around which the center of the sun travels in the course of the year. Also the great circle in which the plane of the earth's orbit, extended, cuts the celestial sphere.

ECTOCRATER: see *Crater.*

EJECTA HYPOTHESIS: the hypothesis that lunar rays, and some other phenomena, are the result of solid bodies ejected from a lunar crater.

ENDOCRATER: see *Crater.*

ENDO-EXPLOSION: an explosion resulting from causes within the moon.

ENDO-LUNAR: from within the moon.

EQUATORIAL PLANE: the plane through the center of mass of any body and perpendicular to the axis of rotation of that body.

EQUATORIAL TELESCOPE: a telescope that is fitted with machinery that causes it to follow a star without readjustment. The rotation of the earth causes all stars to appear to revolve around the earth each day in paths that are parallel to the celestial equator.

EQUIPARTITION OF ENERGY: equal division of energy. A gas acts very much as though it were composed of small elastic bodies. These particles continually strike each other and rebound. Two that have done this will have the same energy as a result. Because of the countless collisions, all the particles in a gas tend to have the same energy of motion. See also *Absolute zero.*

ERG: the metric unit of energy. One erg equals the energy of motion that one gram of matter has when moving one centimeter per second. It is a very small unit.

ESCAPE VELOCITY: the outward velocity needed for an object to escape from a celestial body.

FAULT: a fracture of the surface, along which there has been slippage, either vertical or horizontal.

FILTER: in astronomy, a substance which permits certain wavelengths of light to pass through but which stops all others.

FORESHORTENING: distortion in two-dimensional view of a sphere—objects near the edge appear crowded together in the radial direction.

GALVANOMETER: a device for measuring electric currents.

GEOCENTRIC DIAMETER: the angular diameter of the sun, moon, and planets as measured from the center of the earth.

GHOST: the bare hint which remains of a lunar feature that has been practically destroyed by some later action.

GLOBE-PROJECTION METHOD: a photographic process by which most of the foreshortening of lunar photographs can be corrected. See also *Foreshortening*.

GRABEN: a sunken area between faults. Opposite of *horst*.

HALF-CRATER: a feature that appears to be the remaining half of a partly destroyed crater.

HELIUM: the next to the lightest of the chemical elements (He). It is the end product of the sun's energy-generation process.

HORIZON, TRUE: the great circle in which a plane perpendicular to the plumb line of the point under consideration cuts the celestial sphere. Each point of the surface of the earth or moon has a different true horizon.

HORIZONTAL REFRACTION: the amount that an object which is at the true horizon appears to be raised toward the zenith because of the refraction of light by air. Under standard conditions horizontal refraction equals 36′29″.

HORST: a surface area cut by faults. Opposite of *graben*.

HYDRODYNAMICS: the study of the laws of motion and action of liquids.

IMPACT FORMATION: a crater or other feature that has been formed by impact of a meteorite or asteroid.

INCLINATION: (a) the angle between two planes or, sometimes, the complement of this angle. For example: the inclination of the plane of the moon's orbit to the plane of the ecliptic is 5°08′. (b) Sometimes, when one of the planes is the equator of a body, the context makes it more convenient to tell the same fact as the inclination of the *axis of rotation* toward the second plane. For example: the obliquity of the ecliptic usually is defined as the angle of inclination of the earth's axis of rotation toward the plane of the ecliptic. It is 23°27′ at the present time.

INFRARED: electromagnetic radiation of wavelength too long to be seen by human eyes but too short to be detected by radio equipment. We observe it by means of its heating effect and by use of special photographic plates.

INSOLATION: the total amount of radiative energy from the sun received at any given place during a specified interval of time. It depends on the intensity of radiation and the length of time that it is received.

INVERSE-CUBE LAW: relationship affecting tidal forces, which decrease, approximately, as the cube of the distance of the disturbing body. If our moon were twice as close as it is, the tidal force would be about eight times as great.

IONIZED PARTICLE: an atom or molecule that is electrically charged.

ISOSTASY: the condition under which the weight of a column from the center of a body (such as the earth or moon) weighs the same as every other such column. If the body is not rigid, this condition is imposed on it automatically by its gravitation, its rotation, and any other steady forces. If, however, it is rigid, the strength of its material prevents its conforming to changes in the forces and there may be a lack of isostasy.

ISOTHERM: a locus of points that are at the same temperature. Isotherms are printed on the daily weather maps in many of our newspapers.

KINETIC ENERGY: energy of motion of any body. It depends directly on the mass of the body (doubling the mass doubles the kinetic energy), and on the square of the speed (doubling the speed increases kinetic energy four times). See also *Absolute zero*.

KRYPTON: one of the gaseous elements found to a slight extent in our air (Kr).

LATITUDE: the angular distance measured from the equator along a meridian.

LIBRATION: usually the real or apparent oscillation of the moon that allows us to see some of the hidden side; sometimes the oscillation of other celestial bodies.

LIMB: the "edge" of the disk of the moon, sun, or a planet as it appears from the earth.

LINE OF APSIDES: in the case of the orbit of the moon, the line connecting *perigee* and *apogee*.

LONGITUDE: angular distance measured around the equator. On the moon it is measured from the center of the visible disk when all librations are zero, counterclockwise as seen from the north.

LUNAR TIDE: a very slight bulging of the moon due to the tidal force of the earth on it.

MAGNETIC FIELD: the distribution of force around a magnet. The moon has little or no magnetic field.

MARE (*plural* maria): any of the large, dark areas on the moon or Mars. Because the earliest observers believed such an area to be a body of water, they named it "mare," which is Latin for "sea." Apparently lunar maria are tremendous plains of lava with a thin covering of dust.

MASS: the amount of material that composes a body.

MERIDIAN: any great circle that passes through the poles of rotation of a body. All meridians are perpendicular to the equator of the body.

METEOR: the flaming light observed during the passage of a solid body traveling at high speed through the earth's atmosphere from outer space. See also *Meteoroid*.

METEORITE: a meteoroid that has landed as a solid on earth. See also *Meteoroid*.

METEOROID: a solid body moving in orbit around the sun but too small to be observed from earth.

MICROMETER: a device for measuring very small lengths or angles.

MOLECULE: the smallest particle of any chemical substance. Breaking up the molecule changes the substance.

MOMENT OF INERTIA: a measure of the "resistance" of a body to changes in its rotation.

MOMENTUM: the product of the mass of a body by its velocity.

MOMENTUM, CONSERVATION OF MOMENT OF: when two bodies are moving around their mutual center of mass (for example, the earth and moon), their total moment of momentum is a constant unless some other body interferes.

MOMENTUM, MOMENT OF: the product of the momentum of a body by its distance from any chosen point around it which is desired.

MONTH, ANOMALISTIC: the interval required for the moon to pass from *apogee* (or *perigee*) to the same point again. It is important in distinguishing between predictions of total and annular solar eclipses. Its length is 27.55455 days, slightly longer than the sidereal month, because of a slow forward motion of the apogee.

MONTH, CALENDAR: one of twelve unequal intervals into which the civil year is divided.

MONTH, NODICAL: the interval required for the moon to pass around its orbit from one node to the same node again. It is essential in eclipse predictions because the new, or full, moon must be rather close to a node for any kind of eclipse to occur. Its length of 27.21222 days is slightly less than the sidereal month because of a regression of the nodes around the ecliptic.

MONTH, SIDEREAL: the interval required for the moon to pass around its orbit until it again reaches the same place with respect to the stars, as seen from the earth. This interval amounts to 27.32166 days.

MONTH, SYNODIC: the interval from new moon to new moon. It averages 29.53059 days.

MORPHOLOGY: as applied to the moon, a study of the kinds of features on it, their sizes, causes, structure, and changes.

MOTION: the process of changing place or position. *Linear motion* is the rate (e.g., number of kilometers per second) an object moves with respect to another object chosen as origin. *Angular motion* is the angle on the celestial sphere passed over in some arbitrary unit of time—as, for example, 13° per day eastward around the path of the moon's orbit. *Direct motion* is the eastward direction in which the earth moves around the sun. All planets have direct motion around the sun. However, some satellites move in *retrograde* orbits (from east to west) around their primaries. *Proper motion* is the slow change of a star's position against the stellar background because the star is moving in space. For most stars this motion is only a fraction of a second of arc per century.

MOUNTAIN-WALLED PLAINS: see *Crater plains.*

NADIR: the point on the celestial sphere through which a plumb line would pass if extended downward; it is exactly opposite the *zenith.* Since no two places on the earth have the same plumb line, each location has a different nadir and zenith.

NEWTON'S LAW OF GRAVITATION: each particle of matter in the universe attracts each other particle with a force that is directly proportional to the product of their masses and inversely proportional to the square of the distance between their centers.

NIMBUS (*plural* nimbi): the appearance of a halo or cloud around a feature.

NODE: one of the two opposite points where great circles intersect each other. In astronomy, usually one circle is the *ecliptic* and the other is the circle traced on the celestial sphere by the orbit of the moon, a planet, a comet, or an artificial satellite. The *ascending node* is the point where the object passes the ecliptic from south to north; the *descending node* is the reverse point.

OCCULTATION: the hiding of a body by a much larger one.

OPPOSITION: a position halfway around the celestial sphere from the body of reference. For example, the full moon is in opposition to the sun.

ORBIT: the path of a body around another. The path can be predicted from the law of gravitation.

ORBITAL PLANE: the plane that contains the orbit of an astronomical body.

OUTGASSING: the escape of gas from a solid or liquid.

PANCHROMATIC EMULSION: photographic material that is sensitive to all wavelengths (colors) of light in the visible spectrum.

PALUS: Latin for marsh. A lunar area that looks like a terrestrial marsh.

PARABOLIC MIRROR: a concave mirror having a parabolic surface. Such a mirror will produce far better images of distant objects than a spherical mirror.

PARALLAX: the apparent shift in position of a celestial body against the background of stars, caused by a shift in position of the observer. From the amount of shift, the distance of the celestial body can be calculated.

PERIGEE: the point in the orbit of the moon (or of an artificial satellite) where it is closest to the earth. Opposite of *apogee.*

PHASE (LUNAR OR PLANETARY): as applied to the moon, usually the length of time since the last "new moon" phase. For planets usually the ratio of lighted to total surface as seen from the earth.

PHOTOMETER: any one of various devices for measuring the intensity of a light source.

PHOTOSPHERE: the lowest level that we can see into the sun. It looks like a surface in our telescope and photographs. Most of the sunlight comes from the photosphere.

PLUME RAYS: rays, especially of the Copernican system, that look somewhat like feathers. The "feather" points toward the primary crater.

POLARIZED LIGHT: light waves that have been made to vibrate in a particular direction.

POTENTIAL ENERGY: energy due to position, as contrasted to *kinetic energy.* A weight held above the floor has potential energy. If it is released, the potential energy is transformed into an equal amount of kinetic energy. See also *Kinetic energy.*

PRISM: a transparent solid or liquid (usually glass) bounded by plane sides (usually three) that can change the direction of a light ray or separate a ray into its component wavelengths, giving the band of color called the *spectrum.*

PROTOEARTH: a probable early form of the material that became the earth. According to one hypothesis, the material from which the moon was formed was part of the protoearth.

PROTON: one of the elementary particles that make up atoms.

PROTOPLANET HYPOTHESIS: the hypothesis that the sun was a nebula with a mass greater than it has at present, and that there were various secondary centers of condensation revolving around the center of the nebula. These secondary centers were the protoplanets. See also *Protoearth.*

RADIATION: the form of energy that travels through space at a speed of about 300,000 km. per second. Depending on the wavelengths, it is divided arbitrarily into radio, infrared, light, ultraviolet, x-rays, and gamma rays.

RAY SYSTEM: the bright streaks radiating out from certain of the moon's ringed plains and smaller craters. They become conspicuous under a high sun.

REFLECTING TELESCOPE (REFLECTOR): a telescope in which a concave mirror instead of a lens forms the image of the object to be observed. See also *Refracting telescope*.

REFRACTING TELESCOPE (REFRACTOR): a telescope in which the image of the object to be observed is formed by refraction of light passing through a lens. See also *Reflecting telescope*.

REFRACTION: the change of direction of light as it passes from a transparent medium of one density to one of a different density. Prisms and lenses use refraction to produce the desired results.

RILL: a narrow valley on the moon. See also *rima*.

RIM CRATER: a small crater on the rim of a large craterlike formation.

RIMA: Latin for a rill. Officially adopted by the International Astronomical Union in 1961.

RINGED PLAINS: any of the larger craters of an explosive origin.

ROCHE'S LIMIT: the minimum distance apart that two bodies can exist in stable condition. If they are closer together, the mutual tidal forces will disrupt one or both bodies.

SATELLITE: any astronomical body that revolves around another of much larger mass. The word usually refers to moons, that is, bodies that revolve around the planets.

SAUCER: a saucer-shaped depression in the floor of a mare or of a mountain-walled plain.

SCARP: a cliff or a line of cliffs.

SEAWARD WALL: the part of the wall of a shoreline crater that is toward a mare. Usually it is lower than the remainder of the wall and it may be entirely missing.

SELENOGRAPHIC LONGITUDE: the angular distance measured from the mean center of the moon's visible disk, as seen from the earth, along the moon's equator to the foot of the meridian through the object.

SELENOGRAPHY: the study of the surface of the moon. The lunar equivalent of terrestrial geography.

SELENOLOGY: the general study of the moon and its nature.

SENSOR: a device for detecting radiation. Among the common sensors are photographic plates, radio receivers, bolometers, thermocouples, masers, and the human eye.

SIDEREAL: pertaining to the stars.

SOLAR BATTERY: a device that converts solar radiation directly into electrical energy.

SOLAR DAY: see *Day*.

SOLAR TIDE: the part of the ocean tides which is caused by the sun. The solar tides are a little more than a third of the lunar ones.

SOLAR WIND: the outward streaming of electrified particles from the sun. The winds are continuous although varying in intensity.

SPECTROGRAPH: an instrument, usually attached to a telescope, used to photograph the spectrum.

SPECTRUM: (a) the band of colored light into which a light ray is separated by a prism or a diffraction grating. (b) the series of wavelengths that are emitted by a luminescent gaseous element. No two elements emit the same series.

SPECULAR: mirrorlike.

SPHERICAL ABERRATION: the failure of a spherical mirror to bring to a common focus point all the light from any one point of the light source. This defect is corrected by substitution of a paraboloid mirror for the spherical one. Lenses also have spherical aberration.

SUBSOLAR POINT: the point on the surface of the earth, moon, or a planet for which the center of the sun is at that instant exactly at the zenith.

SUBTEND: literally, "to extend under." Used with respect to the chord of an arc of an angle. If the angular diameter of the moon is 31 minutes of arc, the moon becomes a chord of that angle and we say that the moon subtends an arc of 31 minutes.

SWAN BAND: a series of wavelengths emitted by a molecule composed of two atoms of carbon. It is commonly found in the spectra of comets.

TERMINATOR: the great circle on the moon that is the boundary between day and night. We speak of the sunrise and the sunset terminators. One-half of the circle is always visible from the earth.

THERMOCOUPLE: a very sensitive and accurate electrical device for measuring temperature. It consists of two dissimilar materials (metals or semiconductors) joined together. Heat absorbed by **207** the joint is directly converted into electricity. The temperatures of points on the moon are measured with a thermocouple set at the focus of a telescope.

ULTRAVIOLET: electromagnetic radiation of wavelengths shorter than the eye can see.

VERTICAL DIAMETER: the diameter of the lunar or solar disk which, at a given instant, is in a vertical position.

WALLED PLAINS: see *Crater plains.*

WANE: to become smaller. The visible lunar disk wanes during the interval from full moon to new moon. See also *Wax.*

WAVELENGTH: the distance between successive crests (or troughs) of a wave.

WAX: to grow larger. The visible lunar disk waxes from new moon to full moon. See also *Wane.*

X-RAY RADIATION: electromagnetic radiation that lies within a certain region of wavelengths much less than those of ordinary light.

XENON: a colorless, tasteless, gaseous element that is found in very small quantities in the earth's atmosphere (Xe).

ZENITH: the point where the plumb line, prolonged upward, cuts the celestial sphere. See also *Nadir.*

INDEX

Aspides, line of, 154, 199, 205
Asteroidal impact, and formation of lunar features, 17, 87, 91–92, 103, 104 (Plate 10–22), 105–106, 113, 117, 118 (Plate 12–3), 119–120, 137, 157, 164 (Plate 19–1)
Astronautical convention, 21, 163, 176–181 (Plates 19–25 to 19–30)
Astronauts, see Apollo manned missions
Astronomical convention, 21
Atlas (crater), 30 (Plate 3–15), 37, 52 (Plate 4–11)
Atmosphere, lunar (see also Earth), 6, 9, 11–13, 57, 85
Australe, Mare, 24 (Plate 3–4), 38, 52 (Plate 4–11), 96, 135, 140 (Plates 15–2, 15–3)
Autolycus (crater), 29 (Plate 3–14), 37, 46, 52 (Plate 4–11)
Auzout, Adrien, 11
Azophi (crater), 37, 127, 151 (Plate 16–7)

Bailly (crater), 21, 31 (Plate 3–18), 37, 67
Baldwin, Ralph B., 117
Ball (crater), 37
Barrow (crater), 37
Basalt, lunar, 162
Bay of Rainbows, see Sinus Iridum
Beaumont (crater), 37, 93 (also Plate 10–7), 94 (Plate 10–8)
Beer, Wilhelm, 6, 7, 14
Bessarion (crater), 37, 127, 132
Bessel (crater), 37, 45, 46 (Plate 4–5), 97 (Plate 10–14), 127, 135
Biot (crater), 15, 37, 127
Birt (crater), 37, 90 (Plate 10–3–B)
Birt, W. R., 7
Blagg, Mary A., 7
Blancanus (crater), 37, 110
Blanchinus (crater), 37
Blowhole craters, see Craters
Bond, Dr. W. C., 56
Bond, W. C. (crater), 37
Bonpland (crater), 37, 88 (Plate 10–1), 91, 127
Boscovich (crater), 37
Brayley (crater), 37, 127, 132
Brightness, see Albedo; specific feature
Bright spot on outer wall, 127, 132
Brown, Professor E. W., 152
Bullialdus (crater), 37, 59 (Plate 5–2), 127, 131, 132
Bürg (crater), 38
Byrd (crater), 180 (Plate 19–29)
Byrgius (crater), 33 (Plate 3–21), 38, 55, 61, 116 (Plate 12–1), 127, 135

Caldera, 53, 91, 95, 111, 113, 119, 166, 202
California, central plain, 112 (Plate 11–5), 113
Campbell, W. W., 72
Carpathian Mountains, 38, 106
Carpenter (crater), 38
Cartography, see Maps
Cassini (crater), 38, 52 (Plate 4–11)

Cassini's Bright Spot (crater), 38, 47, 75, 122 (Plate 13–1), 124 (Plate 13–2), 125, 127, 135
Catharina (crater), 38, 108, 127
Caucasian-Alpine-Iridum "island," 105 (Plate 10–23), 106
Caucasian Scarp, 29 (Plate 3–14), 137
Caucasus Mountains, 29 (Plate 3–14), 34 (Plate 3–24), 38, 62, 106, 120, 123, 137
Censorinus (crater), 38
Clarraut, 127
Chappell, J. F., 7, 20
Clavius (crater), 7 (Fig. 5), 23 (Plate 3–2), 31 (Plates 3–17, 3–18), 38, 47 (also Plate 4–6), 62, 67, 78, 80 (Plate 8–5), 108, 110, 112–113, 127, 131, 172
Clefts, see Rills
Clefts (rills) of Hippalus, 62
Cleomedes (crater), 38
Colongitude, 20, 41, 151, 183, 184 (also Fig. 22)
Colony, lunar, 190
Conon (crater), 38
Continental shelves, 42, 90, 93
Coordinates of lunar features, 20 ff
Copernicus (crater), 15, 16, 34 (Plate 3–23), 35 (Plate 3–26), 36 (Plate 3–27), 38, 40 ff, 49 (Plate 4–8), 52 (Plate 4–11), 53, 55, 57, 58 (Plate 5–1), 61, 63 (Plate 6–2), 67–74 (also Plates 7–1 through 7–7), 75, 78, 79, 83, 95, 103 (also Plate 10–21), 105–106, 115, 116, 117, 119, 120, 121, 122, 123, 127, 130, 132 (Fig. 15), 133 (Fig. 16), 134, 139
Copernicus-Kepler-Aristarchus oversystem of rays, 61, 73 (Plate 7–6), 131 ff (also Fig. 14)
Copernicus-Kepler-Olbers-Aristarchus Seleucus oversystem, 61, 111 (Plate 11–4)
Cordilleras (mountain range), 38
Core, lunar, 157, 162
Corona, hydrogen, 195 (Plate 21–6)
Corona, solar, 194 (Plate 21–5)
Coronograph, 194
Crater(s) (see also specific feature; see glossary for types), 14 ff, 17, 53, 55, 60–61, 66 (Plate 6–6), 67, 75, 82 (Plate 8–7), 83, 87, 95, 102 (Plate 10–20), 103 (Plate 10–21), 106, 111–113, 115, 116 (also Plate 12–1), 117–120, 121–122 (also Plate 13–1), 123, 124 (Plate 13–2), 126 ff, 130, 133, 134, 178, 202, 204
Craterlets (see also specific feature), 53, 58 (Plate 5–1), 67, 78, 92, 102 (Plate 10–20), 103 (Plate 10–21), 116–117, 121, 122 (also Plate 13–1), 123, 124 (Plate 13–2), 125, 126–130, 131, 134, 138, 139–140, 144, 149, 164–168
Crater pits, 17
Crisium, Mare, 27 (Plate 3–9), 28 (Plate 3–12), 38, 40, 41–43, 50 (Plate 4–9), 60, 61, 62 (Plate 6–1), 93–95, 94 (Plate 10–9), 96 (Plate 10–11), 103, 108, 113, 135, 149, 157
Crust, lunar, 162
Curtius (crater), 38
Cuvier (crater), 38
Cyrillus (crater), 38, 52 (Plate 4–11), 119, 127, 130